An Ethnography of Knowledge

Afrika-Studiecentrum Series

In collaboration with SAVUSA

(South Africa – Vrije Universiteit – Strategic Alliances)

VOLUME 11

An Ethnography of Knowledge

The Production of Knowledge in Mupfurudzi Resettlement Scheme, Zimbabwe

By

Netsayi Noris Mudege

BRILL

LEIDEN • BOSTON
2008

Photo cover: Delivering cotton bales to the Cotton Company's depot.
Photo by the author.

This book is printed on acid-free paper.

A C.I.P. record for this book is available from the Library of Congress.

ISSN 1570-9310
ISBN 978 90 04 16168 9

PRINTED IN THE NETHERLANDS

Contents

List of tables, figures, maps and photos

Tables

Figures

Maps

Photos

Acknowledgements

This book was an outcome of the cooperation of different people and organisations. I would like to offer my thanks to the WOTRO scholarship fund without which this work would not have been possible. Many thanks go to Prof. N. Long and Dr. P. Hebinck whom I worked with while in the Netherlands. Prof. Long patiently read my drafts, commented and encouraged me to strive for better. Dr. Hebinck never tired of reading all drafts, giving insightful comments and visited me in the field to make sure I was on track and proceeding as planned. Back home Prof. M.F.C. Bourdillon deserves special mention for reading all my drafts, editing and offering valuable comments, and most of all for having faith in me. I also thank Bill Kinsey for being generous with his panel data on the Mupfurudzi resettlement scheme. Many thanks go to Prof. Geschiere and Prof. Murray and Dr. C. Leeuwis for reading my original thesis and making useful comments.

I also would like to make a special mention of my sample households. It was a long process that required dedication and understanding from all of us. C. Mavheneke, M. Karuru, M. Jumbi, M. Mushaninga, M. Gwati, P. Mademo (Tembo), C. Chari, R. Mutyavaviri, D. Seda, D. Maronje, W. Karidza, J. Ngorima, D. Chenjera and E. Mupandasekwa. Thank you! I would also like to thank Mr. and Mrs. Zvorwadza for providing me with a home and a family whilst in the field, and Mr. E. Chidembo, F. Banda, and D. Bwana of Madziwa Mines for providing me with accommodation during the difficult times.

A special dedication goes to my father and mother who always advised me to study to gain knowledge, but to observe and listen to gain wisdom. I found this advice useful in the course of my study. My only hope is that I managed to gain both knowledge and wisdom.

Abbreviations and acronyms

NGO	Non-Governmental Organisation
CIAT	International Centre for Tropical Agriculture
CIMMYT	International Maize and Wheat Improvement Centre
ZANU (PF)	Zimbabwe African National Union (Patriotic Front)
MDC	Movement for Democratic Change
AGRITEX	Agriculture and Extension
AFC	Agricultural Finance Cooperation
AGRIBANK	Agricultural Bank
VIDCO	Village Development Committee
IFPRI	International Food Policy Research Institute
ITK	Indigenous Technical Knowledge
TOT	Transfer of Technology
ANC	African National Congress
HYV	High Yielding Varieties of Maize
CAMPFIRE	Communal Areas Programme for indigenous Resources
COTTCO	Cotton Company of Zimbabwe
HIV	Human Immuno Virus
AIDS	Acquired Immuno Deficiency Syndrome
SEEDCO	Seed Company Zimbabwe
GMB	Grain Marketing Board
CMB	Cotton Marketing Board
ZTA	Zimbabwe Tobacco Association
DERUDE	Department of Rural Development
DA	District Administrator
PGR	Plant Genetic Resource Management
LSCF	Large Scale Commercial Farms

Map 0.1 Zimbabwe locating Shamva

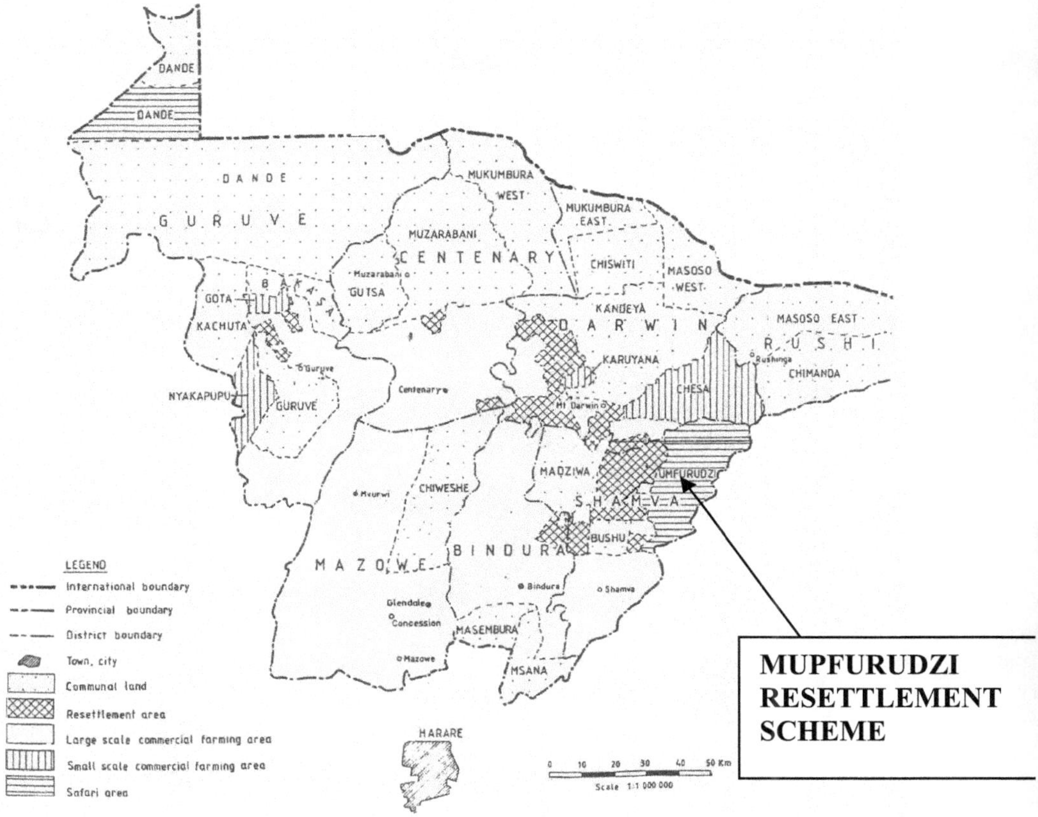

Map 0.2 The location of the Shamva District in Mashonaland Central Province as well as the location of Mupfurudzi resettlement Scheme

1

Knowledge, resettlement and farming

Introduction

This book is based on an ethnographic study carried out among farmers in Mupfurudzi resettlement area in Shamva Zimbabwe. I spent a period of 30 months gathering data. In 2001, I was involved in an externally funded multi-disciplinary study assessing the impact of agricultural research on poverty reduction with a particular focus on High Yielding Varieties (HYVs) of Maize in Zimbabwe (Bourdillon *et al.* 2002). This multi-disciplinary study looked at the pathways of dissemination of knowledge about hybrid maize. The study took advantage of the huge database with quantitative information that was available from previous studies in the same community. The research contained data on 424 households in three land resettlement areas in Zimbabwe. This panel study, unique for Africa, contained data for the years 1984, 1987, 1992, 1993, 1994, 1995, 1996, 1997 and 2000 and covered aspects such as family composition, labour, agriculture, assets, institutional linkages, sources of income, nutritional status and anthropometrics. The database was then used as a reference point from which to select cases for further in-depth studies. As the study progressed and I was confronted with situations in the field, I decided that there was a need to go beyond this rather narrow angle of study to look at the production, growth and dissemination of knowledge about farming in general and not just focus on maize cultivation as a poverty reduction strategy.

This study contributes to academic debates on knowledge. First, since the aim was to investigate how knowledge is produced and socialised, a resettlement area with people resettling from different agro-ecological regions with different knowledge and approaches to agriculture and farming provided a fascinating area. Because farmers were coming into a new area, and were confronted with new crops and new animal and crop diseases in an unfamiliar environment, investigating how farmers negotiated and adapted to this new environment forms an important part of this book. The fact that the resettlement scheme became a melting pot of different knowledge makes the term 'local' a problematic one, yet farmers still use and produce knowledge that is considered 'local'. Second, as shall be discussed later, resettlement aimed to address the racial imbalances regarding land ownership, as well as to improve production among black farmers by resettling them in better agro-ecological zones and by providing them with agricultural experts to help modernise their agriculture. This renders the book relevant to knowledge debates as it unravels how local knowledge makes use of scientifically based state-organised interventions.

This chapter provides a brief introduction to the study, a discussion on the issue of resettlement in Zimbabwe as well as a brief background to the study area. In the discussion on land reform, I will only discuss the early land reforms that occurred in the 1980s soon after independence from British colonial rule, and not the current Fast Track land redistribution. I do this because the study area is a result of the early resettlement and not of the Fast Track Land Resettlement that is too recent for consideration in this study (for a discussion of the Fast Track Land Resettlement see Moyo 2004).

Land reform in Zimbabwe represents a scientific field of great interest. The Zimbabwean government has implemented land reform from above since 1982 in the form of land resettlement. Recently the process of land redistribution has gained a new momentum. The current phase of resettlement involves 'fast track' land resettlement, or land invasions, depending on whether one supports the process or not. A substantial body of knowledge about the process of land reform has already been accumulated (Kinsey 1999) – about asset accumulation, strategies for acquiring income, livelihoods, the effects of land reform on gender and economic empowerment (Gaidzanwa, 1995; Jacobs, 1993) – as well as the social, political and economic justifications for land resettlement (Zinyama 1995: 222). However, little is known about the dynamic processes of acquisition, dissemination and socialisation of agricultural knowledge in the context of land resettlement whereby people move from one place to another, rather unknown, area in terms of agro-ecology infrastructure, institutions and culture.

The lack of academic literature on the issue of knowledge in resettlement areas is hardly surprising, as post-independence academics were mostly interested in evaluating the relative success of resettlement schemes, using the government's stated objectives as a yardstick. For example, there was an interest in whether self-reliance was increasing, whether jobs were being created, incomes improved and food security achieved. Feminist scholars began to focus on issues related to women's livelihoods. In most cases knowledge production was not regarded as an integral component of resettlement since it was assumed that the resettled people were to be 'given' knowledge by the government employed extension workers, and researchers were often concerned that the number of extension workers was insufficient to ensure the effective dissemination of knowledge to the 'ignorant' masses. Only recently in Zimbabwe has there been an attempt to study farmers' knowledge and to question the efficacy of highly standardised expert knowledge (Murwira and Hagmann 1995: 302; Matose and Mukamuri 1993: 28). For a long time there has been an unquestioning acceptance of 'expert knowledge' as the panacea to the problem of low production and poor and inefficient resource use among local farmers. The stress on the paramouncy of expert knowledge (as will be discussed in later chapters) has its roots in the colonial era. 'Official knowledge has a history of being considered as scientific and modern, developed as it was in European centres of knowledge during the colonial era. Farmers' knowledge had little room in the scientifically tested and proven body of knowledge' (Matose and Makamuri 1993: 27) However, in contrast I focus on how farmers in resettlement areas produce as well as internalise knowledge and technology in their lives, and how these processes of internalisation and adaptation of knowledge transform their livelihoods. This study is in line with the growing international interest in farmers' knowledge.[1] This interest is increasing because of the discovery that 'such knowledge is indispensable in view of the need to rebalance growth factors, increased recognition of the significance of diversity in agriculture and changed perceptions about the nature of innovations and the innovation process' (Stuiver *et al.* 2004: 94).

The present study constitutes an attempt to emphasise the farmer as a knower, and therefore to distance myself from the Transfer Of Technology approaches (TOT), which assume that farmers do not know and have to get knowledge from outside (Roth 2001). I depart from Barth's (2002:2) approach, which emphasises that researchers should focus their scrutiny on the distribution of knowledge, especially its absence or presence in particular people and the processes affecting its distribution. From another angle Keesing (1987: 166) maintains that sociology of knowledge must study both the production and the distribution of knowledge. For Keesing, knowledge is diverse and differentiated into layers (ibid.: 162-3); some can get to the inner most layers and others do not. The position of this book is that no one is completely without knowledge but rather that people may know different

---

1 Internationally the interest in everyday forms of knowledge started in the 1980s. Writers such as Knorr-Cetina (1981) were concerned with showing how expert and everyday forms of knowledge related to the production of scientific knowledge in scientific establishments. Chambers (1983) and Richards (1985) picked up this interest in everyday forms of knowledge but took a different route from that taken by Knorr-Cetina who was studying the sociology of science. Chambers and Richards started to emphasise that the knowledge of ordinary people had to be studied and its useful elements used to enrich science. Warren *et al.* (1995) discussed about the cultural dimensions of development in which they emphasised the importance of what they referred to as 'indigenous knowledge systems' in development. Some of the implications of these approaches and also on how these debates have been taken up in the 1990s and 2000s will be discussed in Chapter 2.

things depending on their social positioning and circumstances. There is also a two-way exchange of knowledge and information between those who traditionally regarded as 'knowers', and those who needed to be provided with knowledge.

I adopt the concept of the social production of knowledge (Woolgar 1983: 244) as my central concept because the production of knowledge entails recognition that knowledge is not out there waiting to be used but, like most other commodities, it has to be produced. In several respects social circumstances mediate in the production of knowledge accounts. 'These accounts are to be understood as actively constructed accounts, rather than passively received reflections of an external world, and they are to be understood in terms of the social circumstances which shape their social construction ... accounts are to be viewed as the end product of a process of construction' (ibid.: 244). Thus in this book there is a conscious attempt to show how local farmers are active in the production of knowledge. The notion of 'production' is limited, however, to the extent that it brings to mind the image of factory production where after the necessary steps are taken in the manufacturing process, the end result is a standardised product. On the other hand, as pointed out by Long (1992; 2001: 170-1, 243), Long and Villarreal (1993), van der Ploeg (2003) and Leeuwis (2004: 101), knowledge can never be standardised, and can never be unitary and systematic since it is multilayered and there are multiple realities (Leeuwis 2004: 101). Also its production entails the interaction of different kinds of actors (farmers, researchers, extension officers, NGOs, etc.) and is not linear. As Long and Villarreal maintain, there is no clear distinction between knowledge producers, disseminators and users.

By regarding knowledge as produced, it is also very clear that 'empirical facts by themselves do not determine the facts of knowledge' (Harvey 1981: 95, cited in Woolgar 1983: 245). Farmers themselves select from an array of possibilities and shape their knowledge and practice according to what they think is proper, moral, and relevant to their needs. Thus what determines knowledge are not 'empirical facts' but how these 'facts' are understood and interpreted by the various actors.

There is also an obsession from the side of experts to understand why farmers do not do as they are told. This reflects, as discussed in later chapters, a failure on the part of the experts to realise that knowledge is social and contextual. This failure is short-sighted because research that does not take into account farmers' perspectives usually lacks relevance to farmers' needs, and its results are less likely to be adopted by farmers.

However, recently there has been an attempt by some research centres to include farmers' knowledge and practices when they carry out their research.

For example, for international research centres such as CIMMIT[2] (International Maize and Wheat Improvement Centre) and CIAT (International Centre for Tropical Agriculture),[3] research on farmer knowledge and practice is central to their applied research agendas. These centres involve farmers in their research processes, taking their needs and perspectives seriously throughout. Results from such experiments are usually relevant to the needs of farmers.

The history of agriculture and agricultural knowledge in Zimbabwe

Knowledge during the colonial era: Official approach

During the early colonial period up to the early 1920s, the government was not very concerned with increasing the productivity of African farmers. But, according to Jacobs, 1991: 34), among the Shona prior to 1904 'European agriculture was insignificant and the African peasantry provided the bulk of the food stuffs'. Palmer (1977: 227) describes the 1890 to 1908 period as the era of peasant prosperity in Southern Rhodesia. In the same vein, to show the prosperity of African agriculture during this same period, Phimister (1977: 25) maintains that a report at the turn of the century described Africans as 'agriculturalists ... who do not view the prospect of becoming miners with any enthusiasm. Their present occupation ... pays better and is a more pleas-

2 In Zimbabwe CIMMYT has initiated research on Open-pollinated Varieties (OPVs) of maize, which are more relevant to the needs of resource-poor farmers. As described in Bourdillon *et al.* (2002), 'in this OPV-endeavour CIMMYT constructs more new networks than Seed Co does with regard to hybrid maize. While Seed Co's networks are entrenched in markets and money, CIMMYT looks for strategic alliances with farmers, the public sector, private seed companies, other elements of the private sector, such as distributors and retailers to select, breed and distribute OPV maize seed. The "*Mother-Baby*" trials in Zimbabwe and the leaflet "*Farmer Voices Heard*" are clear manifestations of this strategy'.

3 However, some international organisations have begun to realise this and as a result their research is more relevant to farmers' needs. For example, CIAT regards farmer knowledge and experiences as important when designing their technologies. In their online CIAT synthesis paper they state that when breeding seed they focus on species that are especially important to the poor people living in marginal environments. For instance, in Ethiopia they developed a bean variety that doubled crop yield even when acres under cultivation were reduced. This bean variety was suitable for conditions of low rainfall that prevail in most parts of Ethiopia, suited local food preparation and had strong market appeal. This seed was very popular among farmers and the farmers named it *roba*, 'pouring rain', dispensing with its commercial name Line A176 (http://www.ciat.cgiar.org).

ant life'. This, however, was viewed with displeasure by the rising white capitalists who wanted a cheap labour force to work in the mines and industries, and by the white farmers who did not want to face competition from black farmers.

Zinyama (1992: 37) notes that, as African agriculture improved, the government put into place legislation that restricted African agriculture, so that whites would not face competition from black cultivators. In 1930, the Land Apportionment Act had successfully divided the land into racial blocks with whites controlling most of the prime land and blacks for the most part stuck with marginally productive land, which later became known as the 'native reserves'. In 1931, the Grain Marketing and Maize Control Act was passed. This Act discriminated against black farmers by facilitating a two-tier pricing policy, which favoured the whites and offered subsidies to white farmers. The major aim of these discriminatory policies was to supply cheap labour to white farms, mines and industries, by making farming non-profitable for blacks. Hence the peasant sector became a producer of labour power rather than of agricultural commodities (Bush and Cliffe 1984). Even at the level of resources set aside to develop agriculture African farming was neglected. Palmer (1977: 244) states that in 1940-1 Africans received £ 14, 107 for the development of agriculture in native areas and reserves whilst £ 208, 207 was provided for European agriculture. In 1945-6 and 1953-4 £ 2 million was spent on African agriculture whilst £ 12 million was spent on European agriculture. These discriminatory policies led to a slump in African agricultural production.

In spite of the various limitations and restrictions faced by Africans, the colonial government regarded African agriculture as conservative, destructive and wasteful. The rural poor were viewed as backward, uncivilised and consequently unknowledgeable (Matose and Mukamuri 1993: 27). As a way to stem environmental degradation in the African areas and to stop urban migration, the government introduced agricultural colleges for the training of the native agricultural extension officers. In 1926, Alvord, an American missionary, was appointed as agriculturalist for the instruction of the native. Alvord was interested in converting people to Christianity through changing their agricultural practices. He wanted to prove to Africans that their agriculture was not effective because of the marriage between African beliefs and agriculture, whereas for Alvord agriculture was practical and not spiritual. Describing his view on Alvord's agricultural policies (in relation to some African religious practices regarding rain making ceremonies in an interview by Sadomba in 1998), Chavhunduka, who had been an extension officer during Alvord's time, was adamant that Alvord believed rain making ceremonies were a waste of time because he did not understand the significance of the ritual. Chavhunduka claimed that even if it was proven beyond any doubt that rain making ceremonies did not result in any rain they helped bring production up. This they achieved by bringing people together for the ceremony, encouraging unity within the village and thus making it easier for people to borrow cattle and other farming implements from fellow villagers as well as to organise work feasts (Sadomba 1999a: 42)

The first serious attempt to Europeanise African agriculture started with Alvord and, as we will see in the next section, his policies and visions have survived the transition from the colonial era to the independence era. Under Alvord two native schools for agricultural demonstrators were set up in Domboshava and Tsholotsho (see Alvord 1958, unpublished). By 1973 southern Rhodesia had one agricultural college that trained Africans for the post of agricultural extension officer (Weinrich 1973).[4]

Alvord explained his advocacy of the demonstrator programme by saying that 'I made the discovery that the African must see things demonstrated on his own level, within his reach, by demonstrators of his own black colour and kinky wool hair ... So in June 1921, I evolved a school for agricultural demonstration work for adult Natives' (cited in Sadomba 1999b). In this scheme of things, farmers were not regarded as agents who could actively reason and create knowledge but as people who had to be taught. They had to be forced to see the folly of their farming ways and appreciate the opportunities afforded by adopting 'modern' farming methods. This would be done by providing farmers with the opportunity to see and observe the fields and success of progressive farmers who had abandoned 'traditional' ways and adopted 'modern' farming methods. The progressive farmers were regarded as progressive because of their having been tutored into the modern ways of doing things by experts.

The agricultural extension officers trained under Alvord were responsible for enforcing laws to protect the environment, such as those requiring people to construct water channels and contour ridges, which were said to protect the soil from erosion as well as to ensure land consolidation.[5] Because of their duties, the extension officers became popularly known as *Anamadhunduru/Madhumeni* after the *madhunduru* ('contour ridges') they were forcing

---

[4] The agricultural demonstrator was someone who, after training, went back to the rural areas to work on his land. His field would be like a demonstration plot for other Africans to see that good agriculture had nothing to do with the use of magic but with the adoption of effective modern agricultural methods. An agricultural extension worker is a government worker who goes around as agricultural advisor advising farmers on better methods and showing them how to farm properly, but who does not actually work any land himself like a demonstrator does.

[5] Land consolidation was linked to the individualisation of tenure for Africans. Tenure had to be individulised and land parcels registered so as to avoid fragmentation of land parcels to different individuals thereby compromising productivity. One justification for changing land tenure was to encourage investment in land, which was not encouraged by shifting agriculture.

people to construct. Musoni (1999) maintains that these policies failed to achieve the desired results as people were afraid that these measures were being made to improve tax collection by the government, and therefore resisted them. 'Colonial conservation policy in Southern Africa often fuelled rural anti-state sentiments, provoking peasant resistance' (Moore 1998: 381). According to Yudelman (1964: 116), from 1941 onwards the government began to use legal sanction – compulsion rather than persuasion – as a means of improving production methods through the creation of the Natural Resource Board. At the same time, farmers resisted the contour ridges that had been adopted from the American model of soil conservation. Their argument was that poorly constructed contour ridges were more susceptible to bursting and concentrated erosion at the end of the contour. This could accelerate gully erosion to levels exceeding those of land that was not protected in this way.

The native extension officers were also responsible for training African Master Farmers. In the early days successful farmers would be identified and given Master Farmer badges, but from the 1960s onwards emphasis shifted to Master Farmer training, where farmers would attend lessons and sit for examinations. In 1960 there were an estimated 9,000 (Daneel 1971: 62) Master Farmers in the country and in 1980 this number had risen to 40,000 (Bolding 2004: 84). These figures included those who had received certificates after training on experimental farms or who had demonstrated their ability to farm well under the agricultural demonstrators. However, the majority of those certified after 1960, were given the certificates after attending oral lessons and writing examinations. Thus after 1960 the emphasis was no longer on the practice of agriculture but on the ability to demonstrate theoretical farming knowledge through taking exams (either oral or written ones) and answering questions to the satisfaction of the extension officer. The Land Apportionment Act led to the creation of Native Purchase areas where a Master Farmer could purchase up to 200 acres of land. Thus those farmers who gained Master Farmer certificates by attending courses offered by the extension officers could be eligible for purchasing farms in the Native Purchase Areas. The purchase areas were meant to compensate Africans for the loss of their right to purchase land anywhere in the country. In the 1970s, writers like Weinrich, who did research among Africans in the Guruuswa purchase areas in Masvingo, wrote about meetings with *Master Farmers and their wives* [my emphasis]. Indeed, in the purchase areas women were excluded from owning land as it was mostly men who received the Master Farmer training that was a prerequisite to obtaining land in the purchase areas.

Despite the rigorous training the Master Farmers received, in earlier publications Weinrich indicated that they were not performing any better than other farmers who they had left in the native areas. Shutt (1997: 555) maintains that the low productivity was a result of the fact that much of the purchase area land was of poor quality, often in isolated areas of the country far removed from transportation lines and the market. However, most people writing about African colonial agriculture associated knowledge and good farming with the acquisition of modern methods of farming, so that any perceived failure in African farming was explained in terms of lack of knowledge. Trying to explain the low productivity of farmers, Daneel (1971: 62) concluded that 'in 1960 it was estimated that about 70% of the African producers had not yet made use of improved agricultural techniques, a factor which contributes towards the low yields per acre'. Indeed, it was not only the colonial writers but even the governments of the day who believed that the acquisition of modern technology made possible by getting the Master Farmer certificate was the end of African farming ignorance and the solution to the problem of low productivity.

The colonial government viewed farmers as children who had to undergo intensive farmer training programmes for them to achieve the status of adulthood. For example, in their discussion of the collapse of the Nyamaropa irrigation scheme, Manzungu *et al.* (1996) argue that there was a breakdown of communication between the state and the farmers. They argue that the government started to interfere in the management of the crops. First, the plot holders were forced to grow cash crops, and then to practise compulsory crop rotation. The irrigation staff appointed by government dictated what plants to plant, the planting dates and the type of seed. Farmers complained and resisted.

As suggested earlier, the official approach during the colonial era varied from period to period. During the early days of colonialism up to about the 1920s, African agriculture was left to develop on its own accord. After 1908, the white government took measures to develop white agriculture and simply neglected African Agriculture. European farmers wooed from Britain and South Africa were offered agricultural training, received bank loans to establish themselves firmly in agriculture and could easily gain access to extension services while African farmers did not get any assistance from the government (Palmer 1977: 243). The Department of Native Agriculture was established in 1926 and E.D. Alvord, an American Methodist Missionary, was appointed as the agriculturalist for the instruction of the natives. Later the government felt the need to curb African agriculture through restrictive legislation.

In the 1930s, agricultural extension officers were to instruct farmers on conservation-oriented agriculture. The government also introduced compulsory de-stocking. At the time the white government found it in their best interest to blame the collapse of African agriculture on the ignorance of the African and did not regard its own policies as a contributory factor. Palmer (1971: 244) neatly summarises the effects of the discriminatory policies on agriculture in southern Rhodesia by stressing that European prosperity after 1945 'was achieved ... as a direct result of African poverty'. In the 1950s, the

Chief Native Commissioners were charged with the development of native reserves so as to increase their carrying capacity to reduce the need for acquiring more land for native occupation. According to the government, all these were policies aimed to inculcate in people a sense of responsibility. However, the success of these policies was limited since people resisted them. As noted by Vivian (1994: 181), 'rural anti-colonial struggles coincided with the period of the government's heightened commitment to agricultural development'. Thus as the population pressure increased in the rural areas, the colonial government felt compelled to improve agricultural performance among the natives by teaching them good farming practices. 'The dominant theme of Rhodesian agricultural history is surely the triumph of European over African farmers' (Palmer 1971: 221).

Key elements of the official approach to knowledge during the colonial era can be summarised thus: legislative measures to protect white farmers against black competition, and a strong reliance on government-trained extension officers to equip farmers with knowledge and a barrage of legislation to force farmers to comply. In theory this policy was contradictory because, on the one hand, the white government realised the need to improve African agriculture, whilst on the other hand, it imposed more restrictions on the African farmer. However, if this is viewed as part of colonial discourse in which the problems of African productivity were ascribed to an embarrassingly excessive lack of farming knowledge instead of seen as a glaring outcome of the unequal distribution of resources, the contradiction disappears.

Post-independence era: Official approach

After independence, the first priority of the post-independence government was to remove all legislation that was felt to be restrictive towards the development of black agriculture. Apart from a paper policy shift and an attempt by government to resettle people into better agricultural zones, not much changed. The post-independence government continued with colonial models. Official knowledge was still considered superior to farmers', or villagers', knowledge (Matose and Mukamuri 1993: 37). For example, government continued with the rationalisation of land use. Also the dissemination of knowledge to the largely illiterate rural masses was still seen as the role of government and carried out through its department of Agricultural Extension Services (AGRITEX), now Agricultural Research and Extension (AREX). Indeed one of the prides of the post-independence government was its increase in the ratio of extension worker-to-people from 1:1600 during the colonial period to 1:800 after independence (SAFIRE 2002: 4), and Mutangadura (1997: 35) puts the post-independence ratio of extension officer to farmer at 1:758.

Many agricultural colleges were established for the training of Agricultural Extension workers to enable extensive coverage. Zimbabwe now has six agricultural training institutions, of which two train students up to diploma level, four up to certificate of agriculture level, two are technical colleges and fourteen are major youth training centres that provide agricultural training. The University of Zimbabwe as well as Africa University (a privately funded University) offer agricultural degrees. Since under this model knowledge is still regarded as that which emerges from 'scientific' approaches, AREX does not have mechanisms to initiate or to assist innovative farmers since scientific institutions are regarded as the most important originators of knowledge and AREX as the most important disseminator because it deals with the farmer directly. Thus, discussing the case of an innovative farmer, Murwira *et al.* (2001: 302) show how a farmer in Zvishavane (Zimbabwe), who started his own practices to reduce soil erosion and improve moisture conservation, was regarded as a 'mad person whose ideas should never be emulated by anyone sane'.

The political context in which post-independence discourses on knowledge took shape needs to be understood. Official knowledge discourses are still highly linked to the politics of land. Spierenburg (2004: 5) correctly points out that

> ... over time, a specific narrative has been constructed to guide and justify land reforms or 'rationalisation' of local land use practices: the 'land degradation narrative'. This narrative has its roots in the colonial period and in Rhodesia/Zimbabwe served to redefine a political problem – the shortage of land in the Tribal Trust Lands or Communal Areas – as a technical problem, i.e. the lack of knowledge of local farmers concerning 'proper', 'scientific' farming methods.

So, in 1980, the government redefined the problem of low productivity among black farmers not as the result of a lack of knowledge but caused by a lack of access to quality land and supporting infrastructure, and therefore embarked on land resettlement programmes while at the same time continuing to invest in scientifically trained extension workers.

However, by the mid-1980s the government started debating (Werner *et al.* 1985: 252) whether or not to 'leave the predominantly white large scale farms relatively untouched as it was argued that government land reform would have a negative impact on national farm output and marketable surpluses'. The government had also realised that it could not meet its target of resettling 162,000 people in the time frame it had set for itself, and so it began to pursue 'a dual strategy of stimulating peasant production whilst maintaining the productive capacity of the commercial sector' (Bratton 1985: 181). It was also at this time that peasant farmers began to be regarded as ignorant to such an extent that they had to be taught farming knowledge by government officers. There was a growing belief that white land was efficiently used both in terms of area used and of yield per unit of land and that, because blacks lacked specialised skills required for the cultivation of crops such as tobacco, production would decline if land was redistributed (Moyo 2000: 7, 14). Thus the political problem of lack of access to land was turned once again into the technical problem of poor access to relevant knowledge.

In the current Fast Track Land Resettlement Programme, government has managed to turn the issues around and explain the low productivity of peasant farmers once again in terms of insufficient access to land, since these farmers are generally regarded as having the knowledge to farm, and the ones who do not have relevant farming knowledge are taught. Thus, even in the resettlement schemes discussed in this chapter, in theory, farmers who did not follow the advice and guidance of government resettlement officers could have their resettlement land repossessed and given to someone else. Moreover, when discussing knowledge farmers may also adopt the official discourses and point out that knowledge agents such as AGRITEX were the most important source of agricultural knowledge and information. This was the case with Mandirozva, although it emerged in the course of the discussion that other forms of knowledge were equally important to her for achieving successful agriculture. Indeed, when asked why he thought one of the study villages was better than another, one man maintained that this was because the villagers had more Master Farmers. For him this denoted a wealth of knowledge in the village.

Just like in the colonial era, the acquisition of knowledge in the post-independence era is still associated with the acquisition of the Master Farmer certificate. Between 1981-1994, AGRITEX managed to train 42,000 ordinary and 8,500 advanced Master Farmers (Bolding 2004: 84), which is an extraordinary feat given the fact that up to 1980 the colonial government had trained a total of only 40,000. Even at the time of this research, villagers in the research villages stated that those villagers with Master Farmer certificates got preference from the District Administrator under the Fast Track Resettlement Programme. This was the case because it was believed at the official level that those with Master Farmer certificates could farm more productively than those without, since they had received the prerequisite training. One young AREX officer said at a field day:

> People should make budgets when they farm. To be able to make these budgets and to do other things as well, we are going to have Master Farmer training programmes for next season. You should attend those meetings to get training. Mr Chidhakwa has his Master Farmer certificate. The certificate has a bull drawn on it. We need those bulls. We want everyone to have them. We are also going to hold a district agricultural show at Chakonda so we need to select those who are going to represent us there.

This Master Farmer certificate alone, as viewed by the AGRITEX officer, could vouch for the farming ability of its owner. Like in the colonial period, the Master Farmers are still mostly male and women can only be wives and daughters of Master Farmers. The reasons for this anomaly are discussed in later chapters. However, some deviations from the norm exist. Zinyama (1992: 51) maintains that in the Save and Buhera districts there was an equal number of men and women in the Master Farmer Clubs, although he did not investigate the dynamics of decision-making and participation in these clubs. In Mupfurudzi resettlement scheme, however, the Master Farmers are all male. Knowledge itself is masculinised and only 'the masters' can have knowledge.

There is, therefore, still some continuity between the colonial and the post-independence government view on knowledge. Local views on knowledge are not entertained in the official discourse. Knowledge is understood to be only that which emanates from the centres of knowledge to be disseminated to the ignorant local 'masses'. For the officials, knowledge is hierarchical and follows rigid channels. Showing displeasure with this state of affairs, Hagmann *et al.* (1997: 3) write that 'the hierarchical one way flow of communication and the low standing of peasant farmers in society, especially as perceived by formally educated bureaucrats, largely prevented their needs from being effectively communicated back into the system'.

According to the Chief of Crops (in the then AGRITEX Department), those farmers who attended farmer training programmes, adopted good modern farming practices and had good relations with AGRITEX, had more knowledge about new developments and ended up having more knowledge and income than other farmers. The bottom line for AGRITEX (AREX) officials is that they are at the centre of the dissemination of knowledge and technology. For them no AGRITEX means no information and no knowledge.

The common man approach to knowledge

There is no one-sided approach to knowledge for the common man. Knowledge is regarded as multifaceted and therefore defies the official approach that believes knowledge is more hierarchical and therefore has to follow proper channels from the top of the hierarchy to the bottom. Although at times the common man can also see knowledge as hierarchical, his conception is more local and often includes elements of counter-hierarchy which can also offer counter-expertise. Different people also have different interpretations of knowledge.

Unlike the official approach, villagers did not regard AGRITEX (AREX) or other state bodies (such as the veterinary services) as the most important disseminators of information and technology. Although respondents maintained that AGRITEX associated more with the 'good farmers',[6] they did not

[6] Note that the concept of 'good farmer' is sometimes used as a technical statement and sometimes from the local farmers' point of view.

attribute the capabilities of these good farmers to their association with AGRITEX officials. This, however, varied somewhat over time. In the early years of resettlement, people said that they had gained all their knowledge from AGRITEX. Therefore association with AGRITEX was seen as essential. In those years, it was regarded as imperative for people to have good relations with AGRITEX because it was the route towards obtaining government resources such as fertiliser and seed packs. In recent years, farmers have begun to feel that there is nothing much to be gained from associating oneself with AGRITEX unless one is thinking of venturing into new crops like tobacco. In 2001, most household heads did not see association with AGRITEX as a harbinger of knowledge but rather as an aspiration to work with the richer farmers rather than with 'poor farmers'. AGRITEX officers agreed with the farmers, but as they saw it, this was not caused by a deliberate shift of policy in favour of the rich, but by a shift in policy emphasis from food crops to cash crops.

For officials, farmers' acquisition of knowledge would mean the end of ignorance and the beginning of material wealth. According to this perspective, poor farmers have a poverty of ideas and no knowledge primarily because they do not adopt official advice. This is where the official view on knowledge diverges from that of most villagers. Farmers do not regard the officials as more knowledgeable than themselves. Some argued that they did not need anything from the officials as they had learned everything they wanted to know from the white men in the commercial farms. According to the farmers, they were not necessarily poor because they did not implement good farming practices, but rather because they did not have access to enough inputs and support services such as financial loans.

On the other hand, farmers did not trust the official agents of the state. During the colonial era farmers had resisted efforts by the state to modernise and develop agriculture as they equated this development with oppression and impoverishment. Coupled with the fact that the farmers also did not like to be told that they did not have knowledge, they learned to deal with the Agricultural Extension Officers with a certain degree of suspicion and even dislike. This suspicion did not come to an end with the arrival of independence. Even when extension workers told people that certain things would work, the people did not adopt those things readily without having seen them work in other people's fields. Thus agricultural extension workers introduced demonstration plots, where they carried out practical demonstrations for the farmers. In Zvomanyanga (a resettlement village in Mupfurudzi but not part of the study sample), for example, in 2003 the department of AREX had a demonstration plot of seven different varieties of Pannar maize seed, and also portions of beans and soya beans, which they were trying to encourage people to cultivate. After having had nasty experiences with certain seed or crop varieties, people tend to resist attempts to reintroduce the crop:

> What happened with Pannar was that it was given to people as drought relief after a particularly gruelling drought. People were given long season Pannar seed and unfortunately another drought occurred. That was in 1994. From then on Pannar lost popularity. Only a few people who had always cultivated Pannar before the drought stuck with it. However, very recently we had a field day for Pannar seed and people are beginning to like it again. However some people say that they do not like Pannar because weevils easily attack it.

Knowledge has a history. Thus farmers always call upon their existing stock of knowledge before they decide whether to accept certain things or not. However, this knowledge might be based on selective perceptions that provide only partial truths.

Although six household heads maintained that some farmers did not have knowledge, the things they associated with achieving this knowledge were very different from the things identified by the officials. For example, officials regarded Master Farmers as more knowledgeable than other farmers. When asked what they considered important before a person could be awarded the Master Farmer certificate, one AREX officer pointed out that:

> In the past, for one to get the certificate, we had one thing that we prioritised. After going through the necessary training the person had to have an implements shed before being awarded the certificate. Some people say they are good farmers but after the season, they just leave their implements ploughs and yokes outside to just rot. That is not being a good farmer. As a result we considered these things before we could give you the certificate. However these days after passing the exams, which can either be written or oral, the person gets the Master Farmer certificate; but we still tell them that it is a must that they should build these sheds. We just do not have the time any more to inspect the farmers' households.

This differed from the perception of the farmers. One woman who had attended some Master Farmer training programmes failed to convince her husband to build a shed for their farming implements. Nevertheless, local people regarded this farmer as a very good farmer. Even when talking to the AREX officer on a separate occasion, the latter mentioned the farmer who had refused to build the tools shed as a very good farmer.

No farmers in the sample, regardless of poverty levels, admitted that they did not have knowledge; however, some well-off farmers were quick to point out other farmers that they considered lacking in knowledge. Although all the people in the sample pointed out one or two individuals in the village as very knowledgeable, some villagers maintained that no farmer could be said to have no knowledge, but rather that most farmers lacked resources or were just lazy. This is what two farmers had to say:

> I think there is no one who we can say does not have any knowledge. If a person knows that when it rains they have to go to the field, sow their seed and apply their fertiliser, then that person has knowledge. Wanting to teach others and to know what you have. That is knowledge. One should also be able to distinguish their property from that of others. For example, you see that goat: if you can tell whose goat it is, then you have knowledge ...

The second farmer went as far as to distinguish between a good farmer and a knowledgeable person:

> If he beats other farmers in terms of yield then that person is a good farmer. If you know what type of crop variety and what it means, then we can say you have the knowledge.

Thus, when it comes to knowledge, the opinions of experts and farmers differed. The 'knowledge experts' focused more on the technical aspects of knowledge whilst the farmers focused on the social aspects of knowledge. For farmers, if a farmer fulfilled the roles that society expected of him then that farmer had knowledge. For example a farmer who was not stingy with his knowledge but disseminated it to others willingly, and one who respected other people's property, could be regarded as a farmer with knowledge. On the other hand, just like the officials and experts, farmers would evaluate someone's farming ability. If the farmer had better crops than other farmers, or animals like goats and cattle, and if s/he did not steal other people's live-stock, then that farmer did have knowledge, though this was not the only consideration.

According to local farmers, a farmer with knowledge did not necessarily constitute a good farmer.

> Those with knowledge are very progressive farmers. They farm together with their families and you can see their lives improve. Those with no knowledge regress. They can get two bags of maize, two bales of cotton or even nothing.
>
> *Are there people that you can say have knowledge but are not good farmers?*
> Knowledge is to know how to farm and get good yields. Some people just know how to talk. Hee, I know this, hee, I know that. But when you go to their fields they do not do anything. Some claim that they have Master Farmer certificates. Sometimes you can even take what they say, do it in your field, while they do not apply what they know. If it had been you Mudege, who would you say has knowledge? It is me because although I am getting their ideas, they are not doing anything with them.
>
> *That is what you call knowledge?*
> Jah. Knowledge is doing your work on time. Like right now we have ploughed all our fields except for *madhunduru*.[7] The other one we did not plough is next to

---

[7] Contour ridges.

that place you helped us to pick cotton that year. Now we are just waiting for rain. As soon as we receive rain, we are going to plant our seeds. Those who have knowledge but do not do anything, their knowledge is not knowledge at all. It is useless because they cannot use it.

There is a discontinuity between this kind of thinking and the thinking of experts. For example, all good farmers have knowledge because one cannot be a good farmer without the requisite knowledge. The difference between the official approach and the approach of lay people is that the lay people do not equate knowledge with getting bumper harvests, but with accomplishing their basic agricultural tasks such as weeding and ploughing on time. On the other hand, experts associate knowledge not only with performing agricultural tasks on time but also with adopting modern agricultural practices such as the use of herbicides.

When it comes to knowledge, farmers use a different frame of reference from that used by government and other officials. The acquisition of knowledge from officials is not the sole definer of knowledge, but rather how one conducts oneself *vis-à-vis* fellow villagers. For most villagers the hallmark of a knowledgeable farmer is whether he is able to feed his family or not. A crop surplus for sale does not really denote that a farmer has knowledge, as officials would like to argue. For villagers, a crop surplus merely indicates that a farmer is good but not necessarily that he/she has knowledge.

A brief background on land resettlement

The land question is an issue of major economic and political importance in Zimbabwe. Moyo (1996) correctly points out that land 'underpins the economic, social and political lives of the majority of Zimbabweans'. Thus the anger at the gross disparities in land ownership between blacks and whites became the rallying point during Zimbabwe's liberation struggle. As the black population increased, the blacks were no longer able to eke out a living of the generally poor soils, low rainfall and overcrowded conditions of the rural areas. This discontent with regard to land distribution culminated in the liberation struggle from the early 1960s onwards, resulting in independence in 1980 (Chitsike 2003: 2). Although some emerging discourses question the assumption that the land question was the sole or most important issue in the struggle for independence (see Alexander 2003), it was certainly a rallying point during the liberation war. Even during Zimbabwe Rhodesia under Muzorewa,[8] it was realised that there was a great need for removing

[8] Muzorewa came to prominence as a leader of the resistance and agreed to a settlement proposed by Smith and the British Foreign Secretary, Sir Alec Douglas-Home, at the end of 1979. He was elected in the first elections allowing black people to vote in 1979, but lost support when he could not stop the war.

the racial division on land. The Muzorewa government, however, wanted only a limited redistribution of land between large commercial farms and the peasant areas (Bush and Cliffe 1984: 81), largely aimed at silencing or thwarting the liberation movement which was mobilised around the land issue. Only 2 per cent of European land was proposed for redistribution during this era, and this 2 per cent was regarded as inadequate by Africans in the liberation movement.

At independence, '74% of all peasant land was in areas where droughts are frequent and where even normal levels of rainfall are inadequate for intensive crop production' (Herbst 1990: 39). Although others, such as Mushunje (2001: 2), discuss the issue of the skewed land distribution between the blacks and whites upon independence mainly in terms of the amount owned by each, it has long been pointed out by other authors such as Skalnes (1995: 155) that inequalities in land ownership become even more apparent when quality of land is considered, especially taking into account that at independence almost one third of LSCF (Large Scale Commercial Farms) were located in Natural regions I and II, characterised by high rainfall and good soils, whilst less than a tenth of communal area farms were in these areas.

On the other hand, 'two-thirds of the country is relatively infertile and heavily drought-prone (Natural regions IV and V). This is where almost all three quarters of communal farms are found' (Skalnes 1995: 155). As a result, after ZANU (PF) successfully used the land question to garner support from the masses when it came to power in 1980, the ZANU (PF) government saw it fit to immediately deliver some of its promises for equitable redistribution of land in post-independence Zimbabwe. The Mupfurudzi resettlement scheme was one of the earliest of such resettlement schemes set up by government in 1980. The vast majority of farmers in Mupfurudzi settled in 1981.

However, the government failed to meet its intended objective of resettling 162,000 families in the first ten years of independence. For instance, Moyo (2004: 7) claims that between 1980 and 1996 only 70,000 families were resettled, which fell far short of the targeted 162,000 families for resettlement by 1990. Skalnes (1995: 156) estimates the figure of resettled families by 1993 at 55,000 families. Land resettlement was fastest before 1985 when 38,000 families were resettled by 1983 (Jacobs 1990: 170), but began to slow down after that, as government began to focus on rural development initiatives in order to provide infrastructure in the Communal Areas, which supposedly would reduce the need to acquire more land for

---

Other black political parties of the time including ZANU (PF) and ZAPU did not recognise the Zimbabwe-Rhodesia government which Muzorewa led as Prime Minister. Instead they regarded Muzorewa as a puppet leader of the white minority regime.

redistribution. Thus the failure of the government to deliver some of its promises regarding land led to land invasions of Large Scale Commercial Farms (LSCFs) in 2000 by peasants, villagers and war veterans who had fought in Zimbabwe's war for independence from the British.

Even before the land invasions of 2000, Matose (1997: 69) was worried that

> ... the slow progress that has been made in land redistribution has driven some landless and poor people to resort to 'squatting' as a means of gaining access to land for settlement and farming. Forest lands and state lands have been especially vulnerable to squatting and illegal resource use by neighbouring communities resulting in conflicts with forest managers.

The land reform had failed to challenge substantially the basic property regimes that had existed from the colonial era.

The resettlement of the early days was based on the 'willing buyer'/-'willing seller' concept. The black political parties, that is, ZANU (PF), and ZAPU, as well as the British Government of Margaret Thatcher, had agreed at the Lancaster House conference to end the war. Thus farmers could be resettled as families on land that had been abandoned by white farmers during the war of liberation or on land that farmers were willing to sell. As noted by Moyo and Skalnes (1990), in the early years most farmers who were willing to sell were themselves living in marginal areas. They saw this as an opportunity to sell their land and buy farms in prime farming areas from some farmers that were leaving. Mutangadura (1997: 18) also argues that most of the land acquired for resettlement was of poor quality due to the fact that land was sold on a 'willing buyer'/'willing seller' basis. Thus, in those early years of resettlement until 1989 when 'illegal' land occupations begun, Moyo (2004: 7) states that 'Zimbabwe's land reform in terms of the amount, quality, location and cost of land acquired for redistribution was driven by landholders rather than the state or the beneficiaries in accordance with their needs and demands'. Although in 1985 the 'parliament passed a new Land Acquisition Act which allowed the government the right of first refusal on all Large Scale Commercial Farms put up for sale' (Chitsike 2003: 7), the government often lacked the resources to purchase those farms. As a result, the early resettlement farms were located in marginal farming areas near communal areas. Thus the Madziva, Bushu, Chizanga and Nyamaropa communal lands surround the Mupfurudzi resettlement scheme where this study was carried out. Although Mupfurudzi receives a high rainfall of between 750-1000 mm per year, in good years, a characteristic of region II, it is usually classified as region IIb because of its poor soils. After the government restored the powers of the traditional chiefs in 2000, the two villages of this research were placed under the jurisdiction of Chief Nyamaropa. Rukuni (1994) points out that at resettlement the government and some traditional leaders preferred the resettlement areas to be near the communal areas of the people they were resettling for minimum community disruption. However, things did not work out that way because in the resettlement area I worked in, there were not only resettled people from Madziwa communal areas, but also others from as far away as Chimanimani, Karoi, Mutare and Murehwa.

Criteria for selection into these schemes included: being refugees or other persons displaced by war, including extra-territorial refugees, urban refugees and former inhabitants of protected villages; being unemployed; being a landless resident in a communal area or having insufficient land to maintain themselves and their families (Kinsey 1982: 92-113) or being a war veteran (Gunning 2000: 159). To qualify for resettlement a person had to be unemployed, or if not, he had to be willing to give up his urban job and focus on farming full-time (Bush and Cliffe 1984: 87, 88; see also Jacobs 1993: 45). At the time of settlement, the household heads were also supposed to be married or widowed, and aged between 25 and 50. Families selected for resettlement were assigned to these schemes and the consolidated villages within them, on a largely random basis. In the sample, 90 per cent of households settled in the early 1980s had been adversely affected by the war for independence in some form or another. Before being resettled, most (66 per cent) had been peasant farmers with the remainder being landless labourers on commercial farms, or refugees and workers in the rural and urban informal sectors.

Land resettlement was based on Models A, B, C and D. In Model A, resettled households were given 5 hectares of land to be farmed on an individual household basis. Model B involved the formation of cooperatives to manage farms on a cooperative basis. Model C was based on the nucleus of a commercial estate while households had their own individual plots but acted as out-growers. Model D was intended for low rainfall areas in natural regions IV and V and involved the use of ranches for grazing by communal communities. However, model A proved to be the most popular. Jacobs (1990: 170) notes that of the 38,000 families resettled by 1985, approximately 35,000 were in Model A schemes, 2,500 in model B and a small number in one of the other types. Mupfurudzi resettlement scheme fell under Model A.

Families settled in one of these schemes were required to renounce any claim to land elsewhere in Zimbabwe. They were not given ownership of the land on which they were settled, but instead were given permits covering residential and farm plots. In theory these permits could be withdrawn should settlers fail to follow the guidance of government-appointed resettlement officers who taught farmers how to farm and adjudicated in cases of conflict between resettled people. The resettlement officers had the legal power to evict settlers from land (Jacobs 1991: 522). Each household was allocated 5 hectares of arable land for cultivation, with the remaining area in each resettlement site being devoted to communal grazing. In return for this allocation of land, the Zimbabwean government expected male heads of households to rely exclusively on farming for their livelihoods. Until 1992, male household heads were not permitted to work elsewhere, nor could they migrate to cities, leaving their wives to work the plots. The then Deputy Minister of Lands explained the government's position thus: 'We cannot give land to the employed since they will not have time to work that land. At the moment they have a lot of land belonging to the unemployed lying idle' (cited in Jacobs 1983: 45). This was a continuation of the colonial policy whereby the government wanted to separate peasant from proletariat. A person could either be a peasant or a proletarian but never both. However, as shown by the excerpt of the discussion with Snoia from Kamhopo village in chapter three, sometimes people flouted the government's rules and looked for employment, leaving their wives to till the land while they worked in formal wage employment to supplement farm income or even to buy farm inputs and implements. Although this restriction has been relaxed, with male heads being allowed to work off-farm (provided that household farm production is judged satisfactory by local government officials), in the sample, agriculture continues to account for at least 80 per cent of household income in non-drought years. However, as will become apparent in later chapters, sometimes this government requirement that resettled people should stay on the land created problems because some of the resettled people were needy cases that did not have farm equipment, and could not hope to raise enough money to buy the equipment and other inputs since other avenues for generating income such as wage employment were blocked for them.

In addition to its political rationale, the government's other objective in resettling blacks was to improve the standards of living of the largest and poorest sector of the population of Zimbabwe while simultaneously indigenising the economy (Moyo 1998). According to the Zimcord Conference Report of March 1980, the Minister of Lands and Agriculture declared that the land resettlement programme was the starting point towards improving the quality of life of the rural masses. According to the National Report of the Government of Zimbabwe in 1980, the land resettlement programme was also meant to facilitate the entrance of blacks into the mainstream economic activity of the country. Since agriculture was the backbone of the Zimbabwean economy, the government intended resettlement to create a rural farming community that would move from subsistence to commercial production. It worked to provide an enabling environment for sustainable economic growth in the resettlement areas. It provided appropriate infrastructure such as roads to ensure the successful marketing of produce: in Mupfurudzi, there are well-developed road networks. The government also provided housing units, clinics and schools in the resettlement villages to improve the quality of life. Initially, it provided widespread access to agricultural extension services, with virtually all resettled farmers being visited by AGRITEX staff in the early 1980s. However, this has changed since 1990 with pressure on government to cut costs by reducing the number of civil servants as well as, lately, by the demand of the Fast Track resettlement schemes for extension workers.

During the early years of resettlement, loan facilities were made available to farmers through the Agriculture Finance Corporation (AFC) through which they could access cattle, seed and fertiliser packs. However, this again has changed as government has withdrawn from providing loans, and has left this largely to private seed houses such as Seed Co., Agricom, Cargill, and Cottco.

Two further features of these resettlement areas must be underlined. The first phase of resettlement in the 1980s was criticised by social commentators for not being gender-sensitive since it did not target women as a group (Rukuni 1994; Jacobs 1983). Jacobs (1991: 522), for example, was disappointed that the land reform did not challenge gender relations since widowed or divorced women were normally not able to settle in their own right because they were not regarded as household heads. 'Despite the euphemism that the assigned land belongs to the household, settler women lose access to it upon divorce' (*ibid.*: 552). Moyo (2000: 21-22) also recognises the gendered nature of access to land when he comments:

'Patriarchal land tenure value systems among both the white and black community have consistently discriminated against women land owners. In principle therefore more women need to be provided with access to redistributed land in order to achieve a greater gender balance in land ownership, the agrarian structure and in society at large'. It is the nature of resettlement that no women initially had plots registered in their names.

Secondly, the nature of the settlement pattern in these households deserves further comment. Unlike the homestead pattern of settlement found throughout much of rural Zimbabwe, households in these resettlement schemes live in clustered villages (of between 12-60 households), which are relatively far apart. The physical isolation of these villages has precluded the development of small markets there. Indeed, a striking visual feature of these places is the absence of shops or trading areas, since it was not profitable for shop owners to establish shops there. The only shop in Muringamombe, one of the villages in the study, was always well-stocked with alcohol but did not offer any basic commodities, for which people had to travel about twenty kilometres to Madziwa Mine. To access medical services and markets, villagers had to go to Zvomanyanga or Chakonda, a great distance from most villages.

Study villages

Shamva is located to the North of Harare and falls under natural region 2b, characterised by high rainfall. According to Blick (1972), '*Shamva* is a Mashona term indicating wetness from the fact that if you climb the hill (Shamva hill) during the rain season you cannot leave it and avoid a drenching' (see also Logan 1985: 17 for another translation, i.e. 'place of washing').

The major crop in Shamva, from the days of the first white settlers onwards, was maize. According to Blick (1972: 49) cotton was first cultivated in Shamva in the 1920s by the Moubrays. Other farmers took this up spasmodically but soon dropped it since they did not get good yields because insects destroyed most of the crop. Later, cotton became a major crop in the area after the federation when insecticides were introduced and spraying started between 1965-1969. Hence cotton is of particularly recent introduction in the area as compared with maize. And tobacco is a very new crop as well, with some villages in the resettlement scheme having started cultivating tobacco only in the past two to three years (2001-2003).

The soil types in the Mupfurudzi resettlement scheme vary from village to village. There are two main soil types recognised by the villagers. *Shapa* (sandy soil) and *Hova/Chimbangu* (red loam). *Shapa* is said to be good for planting tobacco whilst *Hova* is suitable for cotton and maize. That, in effect, explains why some villages focus mostly on tobacco while others mainly on cotton and maize.

Although villagers acknowledge that these resettlement villages are located in a better agro-ecological region than where they previously lived, some farmers state that whilst they are grateful, they would have preferred to be settled elsewhere where they could have obtained more land and better soils.

Muringamombe, another study village, is located 2.5 km from the main road and is composed of 28 plot-holders, though the number has risen to almost double the original number due to the fragmentation of land within families. In 2001, there was a mass exodus of young men and older single women from the villages to the fast track resettlement farms.[9] However, by

---

9 All the women and a very few young men I knew who went to settle at the farms at the height of farm invasions came back amid allegations that some (mostly male) self-appointed leaders were demanding bribes to allocate land to people. As these people could not afford such bribes they had no option but to return home and again live with their brothers or parents. However, a few lucky ones have managed to stay on at the new farms. In addition, some of those who came back managed to acquire land by applying for the 'Fast Track' land through government channels. Moyo (2004: 23) claims that women benefited more from Fast Track resettlement than under the earlier resettlement schemes. 'Women

the end of 2002 most of these young men and women had returned to the villages amid allegations of corruption in the system of land allocation. The level of education in this village is very low with four of the seven respondents (in the original sample) unable to read and write. Most young people received only basic compulsory education and very few went on to 'O' level.

Mudzinge (popularly known as Chingerengere) is located about 2 km from Madziva Mine. This village is composed of 33 original plot-holders but, like in Muringamombe subdivision, the number of people who lay claim to stands has more than doubled. Six out of the seven respondents (original sample) in this village were able to read and write. Even though resettlement rules stipulated that in order to be resettled people had to renounce wage work, some of the people in this village continued to work at Madziva mine until they were forced out of work when the company closed down. A large number of youth in Mudzinge attended secondary school. This might be related to the levels of education of the first generation of settlers or to its proximity to Madziva Mine and its urbanising influences.

Although these resettlement villages were established in 1981, some people arrived as late as 1983. The time of arrival had an effect on the location of fields with the exception of those who had, for one reason or another, to relocate their fields. Those who arrived early have fields close to the village, whilst the late-comers have fields much further away.

When they first arrived in these villages, the villagers were poor and needed assistance from the government. The government provided help through the provision of loans by the Agriculture Finance Corporation and technical expertise through the Department of Agriculture. Now there is Agribank, a bank that gives loans to farmers, although at the time of this research none of the farmers in the area had accessed any loans. Ever since, people have depended on the provision of loans by the various loaning organisations that now exist in the area. Although some people could manage without loans, such loans are regarded as good sense financially or as a hedge against climatic risk.

Although the villages have always been dependent on government, they have also initiated some home-grown projects. In Muringamombe, there was a gum tree community project and a paddock project.[10] In Mudzinge, there

---

who have been traditionally marginalised in most development programmes, fared somewhat better in A1 land allocations, gaining as individuals an average of between 12 and 24% of the land allocated under the scheme across provinces. In the A2 scheme, women as individuals accessed an average of 5 and 21% of A2 land allocated across the provinces' (Moyo 2004: 23)

10 At the beginning of the study, people talked about the paddock project, which they needed to protect their grazing areas from use by people from nearby villages. Up to now this project has not yet materialised amid allegations of misappropriation of project funds by the leaders. Villagers contributed money for

was a community garden and a field for producing food for the crèche. They also built the crèche, which unfortunately later collapsed due to excessive rains. However, corruption within these projects has threatened their survival.

The sample

As alluded to earlier, this study is an extension of an earlier interdisciplinary study on poverty reduction and high-yielding varieties of maize. For the earlier IFPRI study Mupfurudzi resettlement scheme had been chosen as part of three of the earliest resettlement schemes in post independence Zimbabwe. In this study a sample of fourteen households was used. Purposive stratified sampling was adopted to select the sample. Using existing data from a twenty-year panel study on the early resettlement schemes by Kinsey and others, we were able to identify households and group them according to certain social characteristics that at the time were thought to be important to gain access to some of the different characteristics in the sample population. We[11] grouped the households according to their wealth-ranking (using Kinsey's data on household size, cattle-ownership and maize yield), family size, sex and age of household head, and then selected the households from the resultant sub-groups. This was to ensure that the various characteristics within the sample population were represented in the final sample.

The continued use of the original quantitatively selected sample for the qualitative study that I later pursued is not meant to be a pretension to scientism. The present study was in every sense of the word an extension of research work that had already been carried out in these resettlement schemes. The use of the sample of 14 households that we had used in the study on HYV maize allowed for continuity since I could also draw upon some of the data we had already gathered in the HYV maize study. Furthermore, I managed to create close relationships with these families/ households and was reluctant to start with new ones altogether. Choosing the initial sample from a pre-existing data-set also was a time-saver.

Governance

Following resettlement the Government of Zimbabwe appointed resettlement officers and set up elected Village Development Committees (VIDCOs) to take over the functions of the older, inherited and appointed headmen. VIDCOs were initiated as a way of promoting decentralisation, and as a

---

the purchase of the fence for the paddock but no fence has yet been bought and the money cannot be accounted for.

11 'We' refers to the people who were part of the research team of the IFPRI-funded interdisciplinary research project.

result VIDCOs were part of the local government system. The total area that was traditionally led by one village head now fell under the jurisdiction of one VIDCO. The VIDCOs were mandated with carrying out tasks such as distributing land to new settlers, tasks that had previously been executed by the traditional authorities. Kwangwari (forthcoming, thesis) notes that

> ... the VIDCOs' functions include enabling the villagers to identify and articulate the villagers' needs, the coordination and forwarding of the villagers' needs and proposals to the Ward Development Committee (WADCO). VIDCOs ... are expected to cooperate with government extension workers, operate market stalls, income generating activities, health posts, adult literacy classes and craft and technology industries. The VIDCO includes a youth representative and a women's representative.

In 2000, the government re-introduced the position of headman. Hence in Mupfurudzi, the villages reverted to the institution of headmen, with a separate party structure. The power structure of these villages changed around 2000 with the shift from the VIDCOs (Village Development Committees), which were part of the ruling political party ZANU (PF), to traditional *Sabhukus* (Headmen). The major reason for this change was that local people complained that the VIDCOs were headed by people who came from other areas and did not know enough about local traditions and religion and did not respect them. The local people who complained were mostly local chiefs who were not necessarily in the resettlement schemes but who were now becoming increasingly politically powerful. They maintained that the elected VIDCOs could not communicate with the spiritual guardians of the land, which was seen as causing pestilence and hunger in the land. However, some political commentators, including the political opposition Movement for Democratic Change (MDC), claimed that the move from VIDCOs to Chiefs and village headmen was undertaken for political reasons since the ruling party wanted to use the traditional authorities for its campaigns. Initially in the new dispensation it was understood that the village headman had to be elected. However, after much protest from the local chiefs it was agreed that the village headman had to be chosen from local people within the resettlement scheme who had the same *totem* as that of the royal lineage. Thus in Mupfurudzi resettlement scheme the village headmen were of the *shava* (symbolised by the eland) totem.

This new approach to governance has created extreme tensions between the resettled people and the leadership of the surrounding communal areas. Some resettled people are of the opinion that the traditional chiefs are imposing new laws and regulations that are deliberately harsh in order to harrass the resettled people.

Nevertheless, farmers remain loyal to the governing ZANU (PF) party, which is believed to have helped them in the past. For example, when the government distributed seed packs and fertilisers, and provided loans for building houses following resettlement, these actions were credited to ZANU (PF) and the President. Likewise the party has successfully used local government, chiefs, credit loans and local extension services as a way of extending its control over rural areas. This pattern of crediting the party and not the government is found in other domains. For instance, Natal Common (a groundnut variety) is popularly known as *kaMugabe*, to indicate that people obtained this seed from Mugabe. Members of communities also support the invasion of commercial farms and have themselves participated in land invasions, although, as mentioned earlier, some are disillusioned by their failure to secure land in the fast track resettlements.

The ruling party, then, is a visible actor in the resettlement villages. The party imposes strict rules that control the behaviour and activities of other institutions and their personnel in the area. Agricultural extension officers, teachers and nurses are supposed to support the party views. For instance, in the case of a political party meeting during school hours the school is required by the local party leaders to send a teacher or two to represent all the other teachers at the meeting.

Organisation of the book

This book is arranged into eight chapters. Chapter 1 has shown the diverse backgrounds of people who ended up in the resettlement schemes. Whilst some resettled people had previously been urban workers, others had worked in Commercial Farming areas and yet others had been landless peasants in the communal areas. Because these people came from diverse backgrounds, they also came with differing expectations of what resettlement should achieve for them, and different kinds of knowledge that had to be adapted to the new circumstances in which they found themselves. At the same time they had to forge new friendships and alliances. This chapter has mapped how knowledge (both official and local) emerges from struggles, negotiations and accommodations that take place within a context of a multiplicity of actors, interests and values (cf. Arce and Fisher 2003: 78).

Tracing resettlement and looking at the context within which people were resettled sets the tone of this book. Its main purpose is to understand knowledge within context. To understand what people know, the actors have to be situated. Moreover, the notion that knowledge is socially constructed requires an understanding of the social context within which such construction takes place. The purpose of the present chapter has been to appraise the reader of the context within which the various knowledge discourses operate. I have also sought to highlight the political context of resettlement. Resettlement was, and still is, associated with the ruling party ZANU (PF). This has affected a variety of relationships, especially where knowledge is concerned. It will become apparent that because of its history, knowledge in Mupfurudzi is highly political in nature.

This chapter has also shown that official discourses on knowledge have not shifted much as a result of the movement from the colonial to the post-colonial era. Apart from the removal of restrictive legislation that worked against black farmers to the benefit of white commercial farmers, not much has changed. In both eras, farmers were, and still are, never regarded in any way as originators of knowledge or even as innovators, but rather as adopters, adaptors and rejecters of knowledge. Farmer creativity and knowledge is often overlooked and rarely acknowledged. What farmers know is often juxtaposed with what scientists know and the farmers' knowledge is habitually found wanting. However, as demonstrated in this chapter, people cannot be regarded as being without knowledge, but should instead be seen as having different kinds of knowledge, which might or might not need to be integrated with other knowledge.

The view of the local farmer or peasant farmer as 'ignorant' still persists in official discourses on knowledge. However, of course farmers have not always been vilified as inefficient and destructive, but their position shifts according to the dominant political discourse of the time. Thus, in both the colonial and independence eras, the metaphors and narratives used in official discourse on knowledge have functioned as indicators of the situated selections that the government has made at different points in time to designate who has knowledge and who does not. With this in mind, I agree with Pottier (2003: 7) that 'one cannot discuss knowledge without discussing the economic and political dimensions of its emergence and its use'. Both local knowledge and scientific knowledge are embedded within a specific social, political and economic milieu. That is, the total context within which knowledge discourses are constructed and reconstructed shapes their outcomes. Also, as is apparent in this chapter, farmers cannot be regarded as one homogeneous whole but have different approaches to knowledge depending on their differential placements, interests and abilities within the society at large.

When there is mistrust between the farmers and experts and when farmers are not consulted, decisions by policy-makers can be frustrated at the level of implementation. Knowledge has to gain legitimacy if it is to be regarded as valid. Some policies failed at this implementation during the colonial era due to politically motivated resistance. Sometimes a policy can be rejected not because of a weakness inherent in its knowledge claim, but because of the politics that lie behind the knowledge claim. Knowledge can never be found in its pure pristine 'scientific' form since it is always contaminated by the context within which it emerges. Knowledge is also associated with certain symbols, thus a knowledge-bearer's symbolic capital can determine whether a knowledge claim is legitimised or not. How the knowledge-bearer is perceived is therefore very important. This brings to the fore that knowledge is largely socially produced and based on socially situated selections and network linkages with other local or external actors.

Farmers often strategise in their dealings with official agencies. They may successfully adopt the official discourse for persuing their own personal interests. For example, the extension officers did not understand why farmers still stuck to their 'traditional' ways of doing things after the extension department had pointed them into the right direction. On the other hand, people may seem to acquiesce to the official discourse that designates them as ignorant and attend Master Farmer training whose teachings they may have no interest in adopting, simply in order to gain access to other resources such as fertiliser loans and seed packs. In this way, the Master Farmer certificate was regarded as a prerequisite to access such government resources.

At policy-making levels, the technocrats use formal rationalisations such as measuring production levels, quality of produce, etc., in defining who has knowledge and who does not. Contrary to this, farmers consider the farmer and his environment as a whole. Thus, when decisions are made to use some local farmers as model farmers, knowledge experts should adopt a holistic approach in farmer selection if they are to be effective. This fits in with Pottier's (2003: 5-6) comment that sometimes 'expression of knowledge may say more about the social relations in which they emerge than about knowledge as such'. Pottier (2003) makes the link with the existence of locally varied and disputed technical explanations but here I link this idea with how farmers understand knowledge at the local level in association with social community relations; and also in association with relationships beyond the community level where knowledge can have different meanings for different actors. What the officials regard as 'knowledge' farmers may sometimes regard as 'oppression'.

This chapter has unravelled that the social appropriation of new technology may result in the technology having meanings which are different from or contradictory to the meanings or intentions of the inventors. 'Words can travel, they can be transferred to new social contexts with new meanings and intentions attached to them which have nothing to do with the old semantics' (Probst 1999: 123[12]). The social appropriation of newly disseminated techno-

[12] See also Korovikin 1986 on the different meanings accorded to the birth control pill in a southern Italian community and how these meanings influence the use and non-use of the pill Korovikin (1986) carried out the study in Montebruno, Italy. He noted that the pill was interpreted differently by different people in the community. Unmarried men associated the pill with sexual freedom, so supported its use. Married men associated the pill with sexual freedom and 'were therefore against it, perceived it as undermining any sense of security derived from membership in the family'. For unmarried women, 'to use the pill means to be civilised', thus they used the pill not to prevent pregnancy but to show that they were moving with the times. Thus the pill could be taken occasionally

logies can sometimes hinder or enhance the adoption of new technologies. Thus, as discussed earlier, knowledge can be regarded by its supposed beneficiaries as oppressive and therefore be resisted.

Chapter two discusses theories and concepts relevant to the study, and positions this study in relation to theory. It offers a critique of modernisation approaches to understanding knowledge and advocates constructivist approaches to knowledge. It also discusses and introduces the actor-oriented approach to understanding social phenomena, including knowledge production, and discusses the benefits of studying and understanding knowledge within the context in which it is produced.

The third chapter provides a background with regard to farming in the two resettlement villages, focusing on the type of cops cultivated. There is also a description of how the people came into the resettlement scheme and where they were coming from. Related to this is a discussion of how the people in the resettlement scheme relate to people in the nearby communal areas. The chapter also offers a background to the institutions operating in the area. This chapter sets the tone of this book by exploring the context.

Chapter four discusses how the role of experimentation, observation and popular beliefs shapes the production of knowledge. This chapter recognises that farmers take an active part in the production of knowledge and do not leave everything to the experts. Experimentation does not always end at the factory gate. Farmers' experiments can be assisted by calling on the expertise of the scientist or other knowledge 'experts' or they can be experiments that are carried out by the farmer alone, using his or her own resources. Observation as a central aspect to learning is discussed in this chapter. However, observation and experimentation as well as knowledge are embedded in the social processes that are prevalent in the localities within which they take place. The gendered nature of knowledge is also discussed, as well as the innovation and creativity of farmers in their encounters with adverse conditions. The chapter also challenges the notion of the 'ignorant' and conservative farmer.

Chapter five, titled 'Magic, Witchcraft, Religion and Knowledge', deals with magic, witchcraft and religion and how they affect the production and dissemination of knowledge and information. With respect to religion, the focus is not only on how the different religions facilitate or impede the

---

whilst in the company of friends but its effectiveness in preventing pregnancy was doubted. Korovikin maintains that because of their Catholic background some married women sometimes felt guilty about taking the pill, which was compensated by increasing their participation in the church and confessing their new-found sin to 'our lady of immaculate conception' - supporting their husbands when they pointed out that they should not take the pill because the pill was not good for the health of these women- whilst at the same time taking the pill.

adoption of certain kinds of knowledge but also on how the different religious and magical beliefs affect the social understanding of received information and knowledge, and how this knowledge and information is applied. Not all knowledge received is regarded as fact, but it is analysed and understood in the context of other knowledge. Beliefs *per se* do not impede the adoption of knowledge but affect how people view certain actions and situations. Beliefs in magic and witchcraft are also gendered, with people choosing to behave and act in different ways commensurate with their gender roles. The chapter discusses the knowledge and power axis where sometimes religion can be empowering and sometimes disempowering in situations where knowledge is contested. The chapter also discusses why knowledge sometimes seems self-contradictory in situations where people may believe different things.

In Chapter six, 'Field days: Knowledge Dissemination and Entertainment', I discuss field days as events where knowledge and information are disseminated in a relaxed atmosphere. There are differing perceptions of field days by 'experts' and farmers, and people attend field days for very diverse reasons, some of which might have little or nothing to do with the official reason, that is, the dissemination of knowledge from the experts to the farmers. Field days are also social occasions wherein people are entertained, gossip, and solve disputes, and where power is contested and social hierarchies reinforced. Women, men and children attend field days but they play different roles.

In Chapter seven, 'Knowledge and Practice: Men, Women and Children', I bring to the fore why households can never be discussed as units but should be seen as heterogenous entities in which people may sometimes know different things and engage in practice differently. I look at how men, women, and children use different communication channels and hence know in a different way. I also situate men, women, and children in their different cultural domains in order to understand what they know and what they practice. Decision-making is also analysed in this chapter in order to understand the situated selections that families make and that eventually impact on their agricultural practices.

Chapter eight provides a conclusion. Here I offer a general overview of key issues that have been raised in this book. This chapter will sum up what I have learned concerning knowledge production.

2

Investigating knowledge

Introduction

The production and exchange of agricultural knowledge is an area that has attracted many researchers in the past and up until now. The debate on the sociology of knowledge embraced – and distinguished between – the concept of modern scientific and local (or localised forms of) knowledge (Warren *et al.* 1985; Scoones and Thompson 1993; Richards 1985). It is stressed that both forms and bodies of knowledge represent different, but sometimes overlapping social networks or technological regimes (Wiskerke and Van der Ploeg 2004), and are embedded in different strategic discourses and practices. Many different, and often conflicting, approaches have been used to study knowledge. Notable among these are modernisation, constructivism, Indigenous Technical Knowledge (ITK), Farmer First and Beyond Farmer First approaches.

The modernisation perspective gave rise to the Transfer Of Technology model (TOT) in which knowledge was viewed as only that which could be transferred from scientists to farmers. The major tenet of the modernisation school was that modernisation tendencies would trickle down from the centre to the periphery. The more the periphery was linked with the centre the more knowledgeable it became. At the centre were the experts/scientists who came up with the knowledge to solve the problems of the layman, who had nothing in terms of knowledge to offer the expert. This is the approach that both the colonial and post-colonial governments in Zimbabwe have adhered to. The implications and limitations of this approach to understanding knowledge production and dissemination are themes that will be discussed thoughout this book.

More recently knowledge has been studied as a hybrid phenomenon, expressing the perception that knowledge is neither global nor universal, nor purely local. Long (1996: 47) aptly put this when he wrote that patterns of agricultural development are

> ... subject to the combined effects of globalisation and localisation: that is, local institutions are transformed by becoming part of wider 'global' arenas and

processes, while 'global' dimensions are made meaningful in relation to specific 'local' conditions and through the understandings and strategies of local actors.

This leads to the theoretical positioning that knowledge needs to be perceived as a social construct, that is, as a social relationship (Long 2001) rather than as an artefact or resource that is commoditised and scale- and culture-neutral. For Hebinck and Mango (2004: 286-7) technological packages introduced and favoured by scientists may fail because they 'misunderstand (or misread) and therefore bypass ... culturally embedded notions about agriculture and how to farm'. This perspective of considering knowledge as a social construct positions the approach of this book *vis-à-vis*, for instance, the Sustainable Livelihoods Framework (Carney 1998; Ellis 2000), the ITK (Indigenous Technical Knowledge) and the Transfer Of Technology approaches (TOT) that perceive knowledge as an element of human capital and thus as an asset. This book posits knowledge as the outcome of the interaction between local strategies that people devise to eke out a living, and the political strategies of bureaucratic institutions.

Approaches to knowledge

For the purposes of this book I will only discuss in this section what Long refers to as 'constructivism' in 1996 and as 'constructionism' in 2001. Long[13] (1996: 57) regarded the constructivist perspective as a robust perspective

> ... which provides fresh insights into how 'expert' and everyday forms of knowledge relate to development processes. Such a perspective takes full cognizance of social actors, their values and understandings in the construction of knowledge ...

---

13 In 2001, Long (2001: 244f) prefers the term 'constructionism'. He emphasises that certain types of 'constructivism' tend to emphasise the cognitive and linguistic aspects of knowledge (e.g. Piagetion theory and other perceptual and linguistic theories common in Psychology) to the detriment of the social interactional components. Long's use of the term 'constructionism' instead of 'constructivism' distances him from constructivist theories in psychology or other perceptual theories that are primarily concerned with cognition and linguistic practices instead of broader social practices. Long (2001: 3-4) goes on to state that 'an actor-oriented type of social constructionism, then, requires that we throw our net high and wide. We must encompass not only everyday social practice and language games, but also larger scale institutional frameworks, resource fields, networks of communication and support, collective ideologies, socio-political arenas of struggle and the beliefs and cosmologies that may shape the actors' improvisations, coping behaviours and planned social actions'.

Thus constructivism involves an approach to knowledge that focuses on how knowledge is produced. There is a strong recognition that knowledge is 'socially constructed' (Besbah 2003: 54).

For Mulkay and Knorr-Cetina (1981: 9), an approach to knowledge that is constructivist in nature is

> ... characterised by a concern for the processes by which outcomes are brought about through the mundane transactions of participants. It entails the assumption that outcomes are the result of the participant's interactive and interpretative work. Within this perspective, the sociology of knowledge question of the 'social and existential conditioning of thought' is analysed with a view to the social processes which are constitutive of the production and acceptance of knowledge claims.

Mulkay and Knorr-Cetina's (1981) interaction and interpretation are similar to Long's (2001) practice and interpretation. Constructivism and constructionism recognise that knowledge is socially constructed, and this book takes the same approach. In this book the distinction between constructivism and constructionism is not emphasised and the terms constructivism and constructionism are not used. There is, however, a consistent attempt throughout the book to bring together, analyse and understand practice and interpretation.

The constructivism of the Mulkay/Knorr-Cetina genre arose as an attempt to study 'scientific' knowledge and how science was practised. For sociologists who were not studying how science was made and produced in laboratories, constructivism became a critique of the modernisation school. Constructivist approaches focus not on how facts are preserved but on how knowledge (scientific objects) is produced (Knorr-Cetina 1983). This approach is a result of the realisation that the same thing can be interpreted differently by different actors, resulting in different understandings (Woolgar 1983; for similar arguments see also Cortese 1995). For Arce and Long (1992: 211),

> knowledge is constructive in the sense that it is the result of a great number of decisions and selective incorporation of previous ideas , beliefs and images, but at the same time destructive of other possible frames of conceptualization and understanding. Thus it is not an accumulation of facts but involves ways of construing the world.

To understand knowledge, the attention needs to be on the social context in which knowledge is created and also the processes by which knowledge claims are formulated and strategically asserted. We must investigate the social circumstances out of which selection arises and the resultant knowledge that is produced. For example, Spierenburg (2004: 5-6) shows how a whole body of knowledge was created during the colonial era in Zimbabwe using a land degradation narrative to define the situation in rural areas, thus relegating the shortage of land in the communal areas to a technical problem rather than a to a political one and therefore justifying not redistributing the land to blacks. In the early years of independence, the government rescinded this approach and a whole body of literature came into being which argued that the whole problem was not a technical one but instead a lack of land which led to the resettlement initiative. After a few years of independence, the post-independence government returned to the degradation narrative in order to limit the demands of blacks on white land. Thus the aim of this study is not to show how facts are preserved in scientific statements about nature (Knorr-Cetina 1983: 19) but rather how these facts are construed and read by both 'experts' and farmers to give rise to specific forms of knowledge and knowledge claims.

The above example indicates that knowledge of scientists based on results of experiments is also mediated by the various positions the scientists take with regard to knowledge. In a discussion during a meeting of members of the Journal of the Royal Statistical Society, Fisher (1936: 122-3) said:

> We have always had, and doubtless always shall have, persons who like to speak with authority on experimentation, and whose pleasure it is to take credit for superior knowledge by the simple process of demanding higher precision. If we use ten replications they can ask for twenty; if we use fifty they can ask for hundred. That they say in effect is a good experiment. All you misguided and negligent people are of course content with a lower standard than mine.

Thus even amongst scientists there is no agreement as to what a real experiment should entail.

The main thrust of this book therefore is that knowledge should be understood in its 'social dimension' (Golinsky 1997: 7).[14] The main argument is that there are no objective facts since all facts are painstakingly constructed through a series of selections. Implied in this approach is a critique of classical modernisation approaches, which regard modern knowledge as made of facts and as uncontaminated by the social. If we are to follow Latour's (1993) neatly-laid out argument, science itself is not value-neutral but constructed given the fact that people have never been completely able to separate nature from society. In all societies, instead of purifying tendencies in which nature and society are clearly demarcated, there is instead mediation. Neither nature nor science can be understood independent of each other (Latour 1992; 1993; Drinkwater 1994). Thus in this book there is a strong suggestion that farming knowledge should not be separated from its social and political context.

---

[14] Golinsky (1997) has a big review (and some criticism) of constructivist approaches to science and technology.

Reflections

The constructivist approach has wide implications for the discussions in this book, as well as the actor-oriented approach that was adopted to gather and analyse data. The actor-oriented perspective embodies within it a realisation that knowledge is largely socially produced, based on socially situated selections and network linkages with other local and external actors. As Van der Ploeg (2003: 26) maintains:

> Generalised knowledge as offered by applied science and standardised technologies supplied by agribusiness, both require new forms of local knowledge for their application. As a result, new specific knowledge systems, of a strictly localised character, emerge.

From this perspective the study maps out how knowledge – 'scientific' or otherwise – percolates through to the field and practical level of farming involving both men and women, adults and children, and how scientific knowledge becomes localised and welded with local knowledge.

If we work with the proposition that knowledge is socially constructed and situated in a field of social relations and social life (Knorr-Cetina 1983: 127, 6), then it emerges that at times outcomes are not consciously calculated or even intended by anyone. This puts to shame rational choice theorists who argue that the individual is always calculating the costs and benefits of taking certain actions *vis-à-vis* others. Thus, eventually, this book does not get bogged down in trying to find the 'rational' behind every knowledge claim. Instead, it engages with the more fruitful issue of investigating the ways in which knowledge itself is produced through the various strategies of inter-linking and distancing between actors. This approach also enables me to investigate other social actors who might not be present in face-to-face interactions with the farmers but who are also implicated in knowledge dissemination and production. This is congruent with adopting the position that social life is negotiated and that knowledge entails social relationships. Also to be investigated are explanations involving interests and other factors which are not part of the actors' first-hand experiences.

Throughout I attempt to analyse the social processes and social conditions under which individuals act and make certain decisions. Social structures have a constraining and enabling effect on individual behaviour (Giddens 1976). Individuals might take up what appears to be the most reasonable choice not because they are always consciously carrying out objective evaluations of the available choices and choosing the most profitable one, as rational choice theory would have it. The 'objective' conditions of the social world may set limitations on what is and what is not possible. And in this way, the most improbable practices are excluded as unthinkable since what is possible is already known (Bourdieu 1990: 54; also see Berger 1971 on the social construction of reality and his discussion of typifications). Likewise, for Long (1992: 21) it is essential to recognise that some individual choices are shaped by larger frames of meaning and action (cultural dispositions, or Bourdieu's *habitus* or embodied history) as well as by the distribution of power and resources in the wider arena. For instance, the knowledge of a new hybrid seed in Mupfurudzi (though also subject to subjective conditions) was objectified in institutions such as loan organisations (for example the Grain Marketing Board – GMB). This meant that in spite of individual dispositions people ended up adopting the new varieties and even acknowledged the positive aspects of the new varieties they had not acknowledged when they still had access to the old varieties.

The position of modernisation perspectives (TOT), in which there is a belief in the presence of knowledge in some people, and its absence in others, who needed to be schooled, is not tenable so far as this book is concerned. People are not regarded as having no knowledge but as possessing maybe different sorts of knowledge. 'Knowledge does not have only one way of expressing itself, but manifests itself in a variety of ways' (Haverkort 2003: 12)

Post-modernist tendencies, and globalisation discourses now pervasive in academic debates and research circles, have made it imperative to question the dichotomy between local knowledge and scientific/'universal' knowledge. For instance, local knowledge is often seen as culture-specific and difficult to apply beyond a particular time and setting; while on the other hand, western scientific knowledge is universally valid. This is clearly demonstrated in Schultz's (1964) thesis, where he stated that institutions had to be built to spread relevant knowledge to farmers. For him farmers' knowledge might be relevant but it was poor, not very productive and could not be applied beyond its specific local contexts. In Zimbabwe expert knowledge has always been regarded by state officials and other experts as the panacea to the problems of rural productivity (Matose and Makamuri 1993). In this approach, scientists who are regarded as central to the production of knowledge carry out experiments, replicate them, and after they are satisfied with the results, disseminate their findings for adoption by farmers. Any deviation from this expected order of things is not accepted by experts and government officials. Hence the reason why farmers who experiment are often labelled as 'mad' farmers who should not be imitated (Murwira *et al.* 2001). Thus, following this kind of reasoning, in Zimbabwe schools were set up to train extension workers who, in turn, were tasked to train farmers and even enforce legislation to ensure that farmers adopted good farming and conservation methods (see Sadomba 1999).

However, attempts to implement western technically oriented solutions failed because these solutions did not take cognisance of the imperatives entailed by different socio-cultural contexts, a factor which led to a widespread disenchantment with modernisation perspectives (Hulme and Turner

1990). By implication, therefore, it is 'likely that the so-called technical solutions are as anchored in a specific milieu as any other system of knowledge' (Agrawal 1995: 3; see also Sillitoe 1998 for a similar argument). As shown elsewhere (Sadomba 1992; Bolding 2004), some technologies that were advocated to improve African farming and conservation in Southern Rhodesia (now Zimbabwe), such as contour ridges imported from the American model and destumping (Matose and Mukamuri 1993: 38), led to environmental degradation. It follows from this that if scientific knowledge cannot itself be universalised, then the distinction between local knowledge and scientific knowledge gets blurred.

In response to the above Long and van der Ploeg (2001) have coined the concepts of 'localisation' and 'relocalisation' in which local forms of knowledge are reworked in interaction with changing external and internal conditions. Similarly, Latour (1983: 145) neatly captures the interplay between the internal and external when he talks of 'translation'. According to him, Pasteur and his assistants learned from field conditions,

> ... by translating each item of veterinary science into their own terms so that working on their terms is also working on the field. For instance spore of the bacillus is the translation through which a dormant field can suddenly become infectious even after many years. The 'spore phase' is the translation of the infected field in the farmer's language.

The concepts of localisation and relocalisation are very useful in that they refute the assumption that local knowledge is static/traditional and resistant to change. The concepts indicate the dynamism of knowledge as people blend the old and the new, resulting in knowledge mixes or knowledge hybrids.[15] Thus knowledge is neither purely local nor purely global, which

---

15 Parkin (1995), cited in Arce and Long (2000: 8), in his study of the intertwining of religious and medical knowledge and practice (Islamic and non-Islamic) writes of the rebounding effects of knowledge which he says 'shed light on the complex ways in which specific knowledge practices are constructed and re-transposed or re-accentuated both within and outside the patient/doctor consultation that takes place ...' Unlike structural models of knowledge construction which see this process as an outcome of the interaction between culturally distinct knowledge categories or systems, Parkin (1995) highlights the blending together and the relocation of the origins of belief and behaviour. Parkin here brings out the different meanings attributed to and the different understandings of the healing process by both doctors and patients. He is also interested in how these meanings are reinterpreted both within and outside of the patient/doctor consultation that takes place. This blends in well with the approach taken in this book that knowledge is not static, that it is subject to interpretation and re-interpretation, an approach that rejects the simple distinction between nature and

has led others to use the term 'glocalisation' (Featherstone 1990) or the concept of 'knowledge hybrids' (Hannerz 1990). For Hebinck and Bourdillon (2001: 6), 'knowledge can be perceived as a hybrid phenomenon, neither global nor local. Knowledge becomes localised through a process of redesigning and re-working'.

Knowledge in context

It would be self-defeating to investigate social actors and celebrate their heterogeneous knowledge without ever coming to a conclusion as to how some consensus is formed or at least on how certain forms of knowledge are assimilated by other actors. This indeed has been the weakness of earlier actor-oriented approaches, which were criticised for adopting an extreme form of methodological individualism (Long 1992: 21).[16] For post-structuralists (Foucault 1967) epistemological breaks usher in new eras of knowledge that consequently result in change, yet for others such as Berger (1971) it is the crisis of legitimation that ushers in new knowledge. For Berger, legitimation is the process of explaining and justifying what one knows to others. 'Legitimation not only tells the individual why he should perform not only one action and not another, it also tells him why things are what they are'. The process of legitimation is an exercise of power since it determines what constitutes knowledge and why, at the same time designating those people who are supposed to know. If those in power are no longer able to justify their knowledge it is because they are challenged. The disruptions which result from these challenges lead to the advancement of knowledge as new ideas are incorporated into people's knowledge frameworks. Though the concept of legitimation itself is a useful theoretical tool, it too falls short in that it does not give a proper exegesis of the social actions that lead to old knowledge being challenged and new knowledge being produced in its place, except by saying that when old knowledge no longer applies it is delegitimated.

In order to understand the process of change and consensus formation, Bourdieu (1990) introduced the concept of *habitus* (embodied history) where one cannot predict action from past and present conditions but rather from the continued restructuring of new experience. For Bourdieu (1990: 60-1), habitus tends to ensure its constancy and its defence against change through the selection it makes of new information by rejecting information capable of calling into question its accumulated information, if exposed to it acciden-

---

social. And, as will become apparent later, that agricultural knowledge is intertwined with religious beliefs, magic, witchcraft, technology, health, etc.

16 Long's (1989; 1992; 1996; 2001) actor-oriented approach points to the need for some kind of interface analysis of knowledge generation.

tally or by force, and especially by avoiding exposure to such information. Although this concept of habitus is important in understanding why and how people can cling to old ways, which may or may not be ill-adapted to current conditions, the concept is not user-friendly when it comes to understanding how new knowledge (even if it contradicts the already known, resulting in a problematic situation) is integrated into everyday life.

For this purpose Barnes' (1996: 17) concept of interpretation is useful. More important is the way in which this interpretation is regarded as locally specific and contextual. That is, before people can adopt any new changes they should find the changes both acceptable and useful. 'The local theoretical tradition enters into the identification of genuine facts/genuine phenomenon as distinct from mere artefacts' (Barnes 1996: 28). Arguing in the same vein, Schutz (1964: xli-xlii) speaks of 'provinces of meaning' because for him it is the 'meaning of our experiences and not the ontological structure of the objects which constitutes reality'. An important area discussed in my analysis is how new knowledge (even if it contradicts what is already known, resulting in a problematic situation) is integrated into everyday life, focusing also on how people attach meanings to certain behaviours and practices.

Barnes (1996) also makes a very astute observation, which is a theoretical pointer in our study and informs some of the analysis in this book. For him, there is no valid pathway from 'theory T works', to 'theory T is true', since false theories can make true predictions and false premises can yield true conclusions. During the course of the research, even during the period of writing, no attempt was made to discard any theories that could be said to be false, according to scientific rationality. Even such false theories were often pointers to behaviour that had implications for knowledge and its production. In his study of Baktaman cultivators, Barth (2002: 8) notes that they piled leaves and uprooted vegetation around taro plants because they thought taro liked the smell of rotting vegetation. This is a very interesting case because, even though it could be said that the Baktaman theory was based on a false premise, the actions it produced had real practical implications for knowledge and the cultivation of taro.

Although local farmer knowledge can also be technical, not all local agricultural knowledge is, and not all local farmer perceptions are, technical; some knowledge is symbolic. Technical knowledge focuses on the rational logical aspects of knowledge and symbols refer to things in the human mind which can not simply be reduced to rational encyclopaedic thought or which cannot simply be changed by logical argument (Bourdillon 1990: 343; see also Sperber 1970). Sometimes, the symbolic can also reduce the technical to the symbolic. For example, some people may use technical fertilisers not because they result in bumper crops but because they are a symbol of modern farming although magic and religion might still be regarded as more important than fertilisers. Bourdieu (1990: 54) talks about symbolic capital, maintaining that even scientific knowledge might need to accept this. This symbolic capital is dependent on associations. For example, the acceptance of knowledge might depend on how the knowledge bearer is perceived. Does he symbolise the government at the local level or is he regarded as symbolising something else? Thus, as shown in the following chapters, the book focuses on both the objective reality of technical knowledge and on the agents' perception of this reality.

The relationship between knowledge and power is central. As a way of distancing themselves from the modernisation school, other schools of thought, beginning with the dependency school in the 1960s, came to realise that in most cases knowledge and power are embedded in each other. However, the dependency approach fell short in that it portrayed poor farmers as helpless victims and sidelined their knowledge not because it was unscientific and traditional but because it was a view of the powerless (Sillitoe 2002: 3). There was a need to go beyond the limitations imposed on our worldview by the dependency school. It does not need much persuasion to realise that knowledge and power are always potentially part of each other. Yet despite this realisation, there is no agreement among sociologists on how the knowledge and power axis should be treated. For Parkin (1985: 49), knowledge is not always power while Foucault (Hirst 1985: 182), on the other hand, argues that knowledge produces power (see also Long and Villareal 1994). For Strathern (1985: 65-6) power enters into the dialogues and struggles between persons but it is attached and detached in very different ways.

Although a hasty approach might regard these approaches as different because they seem to point to different conclusions, they are essentially similar in the sense that they all recognise that power is not necessarily centred in certain institutions and certain bodies. For Foucault (Hirst 1985: 174), power has no simple centre 'but it is diffused throughout the whole social body in complex networks and diverse relations.' Turner (1985: 193) accuses Foucault of failing to analyse institutional networks within which discourses are situated. In my view, this is a limited application of Foucault since by analysing the actions of individuals (the smallest social unit) we may also gain access to the understanding of institutional networks within which discourses are situated, since these institutional networks are a result of individual displacement and extension. Foucault's approach has important implications for this book because it recognises that knowledge is always contested and that knowledge is in fact a problematic concept. Thus it can be contested between agents of intervention as well as between agents and other social actors (e.g. NGOs, researchers, agricultural extensionists, etc.). This book demonstrates that the model of state/peasant relationships characterised by a powerful and dominant state on the one hand and a powerless dependent peasantry on the other, is defunct. Power is always contested and negotiated instead of owned.

The status of different forms of knowledge can be affected by a variety of social relationships between bureaucrats and actors in the periphery, and also by the general political situation. Bratton (1980) points out that in Zambia the impact of state intervention in rural areas is not comprehensible without reference to what he describes as the politics of rural development. Knowledge can be regarded as politically sensitive. This is particularly applicable in the volatile current political arena in Zimbabwe. In the past (that is, in the 1980s and early 1990s), in Mupfurudzi, the seed companies were regarded as credible sources of hybrid seeds and technology. However, in the ensuing political tensions precipitated by the 2001 elections, these companies began to be seen as representing the interests of whites, white commercial farmers and the opposition political party (Movement for Democratic Change) who were bent on discrediting the government. Thus the politically powerful people were left to define what knowledge was, and as long as knowledge came from these 'suspicious' sources it had to be treated with caution. To adequately address the power/knowledge axis this book analyses how knowledge is situated in social contexts by looking at individuals within their time-settings. The status of knowledge is always in a flux, and the situation at any one time determines who is more knowledgeable and powerful. The question of the knowledge and power axis is therefore not primarily a question of the use and misuse of knowledge but of how it functions within the system assigning power to some people but not to others (Foucault 1967).

The actor-oriented approach that is used here offers a way out of most of the problems suffered by theories of knowledge. The actor-oriented approach as advanced by Long (2001; 1993; 1992; 1989) does not distinguish evaluatively between different forms of knowledge but regards knowledge as an 'outcome of interaction, negotiations, interfaces and negotiations that take place between different actors and their lifeworlds'. In a subtle way the hybridity of knowledge is recognised. Its explanation of knowledge production and exchange and transformation adopts a dialectic principle whereby the interface is characterised by discontinuity and a critical lack of communication. At the interface level, conflict might arise as a result of differential and often incompatible interests, and on the other hand, interaction, negotiations and strategic accommodations may take place as individual actors try to recruit others into their own 'projects'.

The actor-oriented approach to understanding knowledge is congruent with some post-modernist approaches which regard knowledge as fragmentary and partial (see Jackson 1989; Pool 1995: 25). According to Long (2001) (see also Long and Arce 1992: 212), knowledge is fragmentary, partial and provisional in nature, and people work with a multiplicity of understandings, beliefs and commitments. Implied in this is the fact that knowledge is multi-faceted and contextual. People who believe in one thing in one context may not necessarily believe in the same thing in a different context. By its very nature knowledge is contradictory. Thus, in the chapters that follow I also aim at analysing and assessing the implications of variations in knowledge, social positions, and contexts in local populations. I will investigate how the nature of knowledge itself can facilitate change and the continued production of new knowledge and also how new knowledge can lead to change (e.g. social differentiation).

Knowledge is also tied to beliefs such as beliefs in witchcraft. Recently, there a growing body of literature has come into existance on witchcraft and magic in Africa, or on what is referred to as the 'occult', in relation to questions of 'modernity' (Geschiere 1997; Niehaus 2001; Andersson 2002; Ciekawy and Geschiere 1998). Most of this literature is concerned with showing how increases in witchcraft practice and accusations constitute an attempt by Africans to deal with modernity, and situate the practice of witchcraft and religion strictly within politics (Geshiere 1997: viii-ix). Niehaus (2001) links the witchhunts that occurred in Green Valley, South Africa, with larger political issues linked to the marginalisation of youths from formal politics, since these witchhunts were led by ANC (African National Congress) youths known at the time as 'Comrades'. Niehaus (1998: 23) explicitly makes this link by stating that 'by staging witch-finding rituals, having witches expelled from the villages and by exposing various immoralities the comrades have sought to compensate for their lack of influence in formal political processes'. With the project of modernism in mind, Dolan (2002: 559) chronicles how in a district in Kenya, where contract farming of French beans is highly contested between men and women, witchcraft provided 'a vehicle through gendered struggles over contracts are articulated and contested, and through which the social costs of agrarian transition become apparent' (ibid.).

Many other authors on the modernity of witchcraft have adopted Geschiere's (1997) view of witchcraft as having both levelling and accumulative tendencies. On the one hand, witchcraft 'provides indispensable support for the dominant to accumulate greater wealth and influence. On the other hand witchcraft is a weapon of the weak enabling the poor to sever inequalities'. Thus the rich are denounced as having access to their riches through witchcraft, whilst on the other hand the poor and weak can use witchcraft or threats of witchcraft to gain access to certain resources. However, this approach does not entirely explain those cases where the poor who are regarded as witches are excluded from access to resources by both poor and rich people because of their perceived witchcraft powers. In 2003 there was a panicky rumour that some mysterious poor old woman was going around causing havoc and mischief and that the only way to guard against such a woman was for people to no longer help old women asking for assistance of any kind. As a result some women were beaten up by complete strangers for being suspected of being the evil old witch. Thus being labelled a witch sometimes operates as a discourse of exclusion rather than as a levelling mechanism. The latter views also do not explain why not all rich people, or all poor people, are accused of witchcraft. Although it is not possible to completely reject the levelling or accumulation hypothesis, there is need to go beyond this and recognise that people consider all kinds of evidence before a person is labelled a witch, regardless of his or her wealth or lack of it. Hence one needs to take account of the dynamics and differential interpretations of the significance of witchcraft, not merely of the issue of wealth differences or politics and strategies of exclusion. The latter types of interpretation seem to seek a structural rather than a situational and multiple-meanings explanation, which actor-oriented approaches would stress and which I adopt for analytical purposes.

In the attempt to link witchcraft with modernity there is a desperate attempt to escape from Evans-Pritchard's (1937) view of witchcraft as 'traditional'. But in so doing proponents of this school fall into the same trap as Evans-Pritchard in explaining witchcraft by referring to its perceived functions which are now regarded as serving modern and not traditional processes. Thus despite all the seemingly advanced terminologies used to link witchcraft to modernity, the approach simply inverts Evans-Pritchard's original thesis. In this attempt there is an underlying theme that refuses to disappear, namely that witchcraft beliefs are 'false beliefs' that need to be explained. Rutherford (1999: 92) is also unhappy with an approach that considers witchcraft in relation to modernity and capitalism because it replicates anthropological attempts to know the native 'in terms of western rationality'. In my approach I do not attempt to analyse witchcraft by using modern conceptualisations that relegate beliefs to falsehoods. In fact I do not argue as to the 'falsehoods' or the 'truths' of particular beliefs but rather I approach these belief systems from the point of view of those who believe in them as a reality, whether false or not, that affects various social relationships and the acquisition and generation of knowledge. A recurrent theme in this book is that matters should be understood in their contexts.

> To provide a 'telling' interpretation the ethnographer must find appropriate contexts for elucidating the phenomenon under study. The idea is that in their local contexts matters can be made intelligible, even when they at first glance appear outlandish, exotic, or simply opaque to people encountering them in other times and places (Rosaldo Jnr 1997: 31).

Whereas Geschiere (1997: 13-14) splits hairs trying to justify his use of the term 'occult', stating that witchcraft and sorcery are moralising terminologies concerned with distinguishing good and evil, I am not plagued by the same dilemma. Geschiere uses the term 'occult' to leave open the question of whether witchcraft and sorcery are 'good' or 'bad'. In contrast I have no qualms about using the term witchcraft in the society I studied where witchcraft was considered inherently evil. There was never a time when witchcraft, or what was regarded as witchcraft, was said to be good, and for the people involved witchcraft was also surely a question of morality. However, I introduce the concept of magic into my analysis. I differentiate magic from witchcraft because, as shown in chapter 7, there was a greater ambivalence over whether magic was good or bad.

For Malinowski (cited in Tambiah 1990: 72), 'magic begins where technology ends'. My analysis of the data indicates that this statement should be refuted emphatically. Magic and technology are not separable and are practised jointly. There is no such disjuncture between technologies and magic that one can pinpoint exactly where technology ends and magic starts. For most farmers in the study area they are embedded in each other. Hence I will not attempt to discuss and explore the possibilities offered by discourses of witchcraft, magic and religion to gain control over modern changes, since even before colonisation and its modernisation discourse agriculture was imbued with witchcraft, magic and religion – much to the dismay of most colonial administrators (Sadomba 1999a; Bolding 2004). To show the interconnectedness of religion and agriculture in the pre-colonial days, Sadomba (1999b: 34-6) writes of African agro-religion in Zimbabwe. Borrowing from Geertz (1998: 1), he says that 'the illusion that ethnography is a matter of sorting strange and irregular facts into familiar and orderly categories – this is magic, that is technology – has long since been exploded'.

Magic is a complex issue since, as I observed, some people who denied the existence of magic or at least doubted its ability to work sometimes still took measures to protect their fields from people with bad magic. Even when the wife was the one turning to magic to protect the field, it was impossible to tell exactly whether the men were simply humouring their wives by allowing them to use magic, or whether they somehow believed in magic themselves but loathed to admit to it. Usually men left this duty to their wives, who obtained holy water from prophets and priests. All female-headed households except one had taken measures to protect their crops. Thus, to understand the gendered nature of some knowledge I include a discussion of local beliefs and knowledge in an attempt to unravel their gendered aspects.

Sometimes people's knowledge is based on associations. This is evidenced by the fact that people are not consistent when it comes to knowledge and its application. As a result there is a conscious attempt in this book to understand the context in which people make statements and believe these statements to be correct. For example, one village head in Mupfurudzi stated that people should use fertilisers to get good crops. Yet, on other occasions, he emphasised the power of ancestors to ensure good crops. He was disappointed that some villages had opted out of the rain-making ceremonies and other rituals in honour of the ancestors to ensure bumper harvests. He was convinced that the ancestors would take revenge on these recalcitrant villages. This person was subscribing to two different and seemingly contradictory explanations, depending on the context. In the first context, the village head was a farmer who was struggling with the practical need to ensure good crops and hence subscribed to the scientific theory. In the second, he was acting as the person who interceded between the people and the spirits. If people took recourse to scientific knowledge only, his power based on his ritual capacities would be challenged. Thus knowledge can be regarded as impartial, indeterminate, and sometimes even self-contradicting. It is contextual.

Farmers' knowledge and innovation

Scientists claim that they move from the definition of the problem, sharpening their research tools, and engaging in experimentation and observation, all in search of the 'truth'. Although scientists come up with useful scientific artefacts like in the case of hybrids that can be made resistant to certain pests or diseases, their thinking is linear and can hardly survive in the convoluted world in which these artefacts are supposed to be used. Although the scientists' linear thinking is highly valuable in the laboratory setting, it can be regarded as sterile if it is to be applied by laymen in a world where no variable is controllable. If one is linear in one's thinking then one can be accused of putting on blinkers.

Authors such as Helleiner (1970: 292), who have discussed experimentation in Africa, believed that scientific methods would result in useful products if applied correctly and pointed to scientific research elsewhere that for example had led to improved yields in Asian rice and Mexican wheat. Studies of how technologies fail at the level of implementation are abundant, some asking why farmers do not do as they are told by 'experts', thus focusing on the political, social and economic reasons why farmers reject certain technologies (Matose and Makamuri 1993; Helleiner 1970; Green and Hymer 1966; Monu 1982). Although these studies recognise the fundamental aspect that farmers do things that are beneficial to themselves, farmers are regarded as adopters and rejecters but not as agents who actively carry out experiments to get to the 'truth' as they see it. In this book, I move away from such conceptualizations of experiments and farmers, and focus on the farmer as an experimenter.

Although others like Maurya (1989), Stuiver *et al.* (2004: 104) have recognised the innovativeness of farmers, arguing that 'farmers tend to generate knowledge from practical experiences, and not from formal experiments'. This is a commendable departure from viewing farmers as traditional and very conservative people who do not like change even if it is beneficial to them. However, this approach falls short in that it does not recognise farmers as agents who can actively experiment with a desire to learn and not simply learn from doing what they have always done, what Stuiver *et al.* (ibid.: 106) refer to as an ex-post reconstruction of experiments. Ex-post reconstruction of experiments refers to a situation where experiences are reconstructed as experiments in retrospect.

Some of the data in this book agree with the assertions of Stuiver *et al.* (2004) that sometimes what farmers refer to as experiments are in fact ex-*post facto* rationalisations, and that experiments can happen accidentally, or as a result of improvisation because of the unavailability of inputs, and that usually independent variables are not controlled for. However I differ from Stuiver's *et al.* (ibid.) assertion that it is better perhaps to speak of 'farmers' experimental activities' than of farmers' experiments. For Stuiver *et al.* (*ibid.*) the term farmers' experiments suggests a degree of deliberateness and demarcation that they think is misleading when describing what farmers do. If this assertion is taken to its logical conclusion it implies an attempt not to take too seriously farmers' experiments, albeit unintentionally on the part of the authors. For instance, in Zimbabwe Murwira *et al.* (2001: 302) note that

> ... the government staff regarded any type of knowledge that was locally developed, i.e. that did not find its origins in either the DRSS (Development Research and Specialist Services) and AGRITEX, to be traditional and primitive and therefore not to be encouraged.

In another discussion on *vlei* cultivation Scoones and Cousins (1989) note that although official discourses discouraged *vlei* cultivation if properly managed *vleis* had arable production potentials, and could offer animal grazing at certain times of the year. Local farmers had knowledge about proper *vlei* use. For the government staff technologies and knowledge had to be tested and proven presumably by scientific experiments, not by farmers 'experimental activities'. Thus, in this book, what Stuiver *et al.* (2004) tentatively suggest should rather be called farmers' experimental activities, I call farmers' experiments. Scoones (1993: 14) notes that 'studies that explore the dynamics of farmer experimentation also show that rural people empirically examine alternatives leading to progressive learning'. In this book both scientists' experiments and farmers' experiments are understood in terms of the social circumstances that shape their social construction. Results from farmers' experiments sometimes have more effect on what the farmer does or is willing to do regardless of the existence of conflicting or collaborative results from 'scientific' experiments. Also, given the scientific dilemma that seemingly 'scientific' results can be contradicted by other seemingly 'scientific' experiments, who then is to say that the experiments of scientists are 'experiments' to a greater extent than farmer's experiments are?

The issue of observation (see chapter 4 and chapter 6) is neither a new issue nor an issue that is uniquely Zimbabwean or African. It emanated from the assumption that if farmers saw the miracle of modern farming methods they would believe in and adopt modern ways. In 1892, Bolley (p. 270) wrote that 'the average farmer is eminently conservative when asked about his routine of work. He dislikes innovations as to methods and distrusts ways and means not clearly practical' (see also Hedrick 1918). In the 1900s, the need to disseminate practical information to farmers in America led to a debate on the usefulness of demonstration to give rise to a new breed of farmers that Hedrick (*ibid.*) referred to as the tutored farmer. 'The farmer reacts to no other educational stimuli so quickly as through being shown the successful achievement of some neighbour farmer. 'Pick up in one place the instance of a successful farm achievement by one farmer and carry it to the farmer in other places', says an experienced demonstrator, 'you will win their confidence and adherence at once' (*ibid.*: 162). This was the start of the extension officers and agricultural demonstrations programme that Alvord later imported into southern Rhodesia.

The research and the book

The book looks at the dissemination of various types of knowledge and technologies introduced from 'outside': starter packages of hybrid maize seed and fertiliser, recent introductions of new hybrid maize seed such as Sc501, 502, and other related technological agricultural advice, provided by outside agents such as extension officers and other interested organisations. These are not viewed as distinct bodies of knowledge separate from local knowledge and theoretical frameworks, but as knowledge that is localised and relocalised into the locally specific contexts of individuals and communities. Consideration is given not only to how these new technologies are adopted but also to how people rework them and give them new meanings, which may or may not have been intended at the initial dissemination. As noted in later chapters, farmers employ various linking and de-linking strategies when they come into contact with various agents of change. These actions by farmers affect the production and reproduction of agricultural knowledge in general. Thus the book attempts to unravel the struggles, negotiations, contestations and accommodations that take place between actors during the production of knowledge – a production which is primarily social.

A further issue to be discussed is the dissemination of new knowledge across generations, both formally through the school system and informally through socialisation in agricultural work. The research explores the lack of coherence between these two systems in relation to agricultural knowledge and practice. Related to this is the need for a rigorous analysis of communication.

> Culture ... is a series of communicative acts, and differences in the mode of communication are often as important as differences in the mode of production, for they involve developments in the storing, analysis and creation of human knowledge as well as relationships between individuals involved (Goody 1977: 37).

Here Goody is interested in accounting for the differences between literate and illiterate societies in terms of development and ideas. The present study, in contrast, is not concerned with outlining the differences between literate and non-literate societies but with how knowledge is accumulated and passed down to the next generation within society (for more on communication issues see also Mundy and Compton 1995). Communication leads to the continuity and spread of knowledge within a society. But do the differences in the knowledge that people have in the same society have much to do with the communication channels that they use, or even with levels of literacy? For example, as explored in Chapter 7, the young preferred the written word while the old and women preferred oral communication from a trusted friend, or knowledge acquired in a practical manner. Mombeshora (1994) documents how the spread of literacy led to the abandonment of traditional beliefs on fertility. Change was mixed with continuity whereby, although literate people were able to articulate their scepticism and denounce witches and lineage rituals, they remained afraid of curses. This raises the question of whether this is merely a result of literacy, or if it vindicates Jackson's (1989) assertion that indigenous knowledge is grounded in certain cultural assumptions. Thus, in looking at how knowledge is fabricated, and accounting for differences in knowledge across gender or even intergenerational differences, these things need to be looked at critically.

The study also focuses on local knowledge, including belief systems that impact on agricultural practices. It has been suggested that the best way for development is through reliance on traditional spirituality and cultural knowledge (Chivaura and Mararike 1998). Past fieldwork experiences in the research area have encountered incidences of traditional beliefs in witchcraft affecting a variety of relationships. The focus on spirituality and cultural knowledge was necessitated by Fairhead's (1993: 199) astute observation that

> ... the focus on technical knowledge isolates agriculture from the social context, or put another way the farmer from the person. Researchers who are permitted to examine agriculture in terms of agricultural knowledge can maintain themselves in ignorance of the multitude of non-agricultural influences which inform agricultural practices.

For example, in Mupfurudzi there were incidents of traditional beliefs in witchcraft affecting a variety of relationships and activities. Although anyone could be accused of witchcraft, the people who were accused of having this magic were usually those people who were regarded as 'good' farmers. Focus was not on the falsity or truth of these beliefs but on how they impacted on agricultural knowledge and informed agricultural practice in general. Given the fact that 'traditions' are not static, there is a recurrent theme concerning how these traditions were continuously invented and reinvented and how this inevitably impacted on knowledge.

The third area addressed is the interplay between practical experience and the cultural and localised meanings of knowledge. This entails bridging the gap between 'outside' and 'local' modes of knowledge. Farmer knowledge is dynamic. As noted by Fairhead (1993: 193), locally, nobody is in a position to say what is right or what is wrong and to turn a farmer's hypothesis into truth. For him 'local knowledge lies as much in its methods, in its lack of overbearing authority and in its fluidity as in what is known'. Linked to this is the issue of how farmers develop the confidence to follow their own initiatives and come up with new forms of knowledge. The book offers an attempt to understanding how farming and farming knowledge is embedded in the social and political life of the actors. Linked to this is an analysis of the networks that actors utilise and how they impact on their own interpretations and understandings. This theme is touched upon at various junctures throughout the book.

I adopt a critical stance towards local knowledge because such knowledge cannot be a panacea for the problems of agricultural development. A close reading of the coming chapters will show that one cannot conflate local knowledge with good farming practices and neither should one attempt to judge local people as ignorant because of the knowledge frames they adopt. Apparently, even farmers are 'objectivists' in the last instance because knowledge has to deliver and farmers judge it based on its efficacy. Hence I look closely at the role of experimentation and local observation in the acceptance, rejection or adaptation of new technologies. The issue of adaptation or reworking of new technologies has recently attracted the attention of agricultural scientists. 'Adaptation' and 'reworking' have recently been understood as the production of (often hidden) novelties (Wiskerke and Van der Ploeg 2004). In the case of technical knowledge, there is room for the correction of error through experimentation and reworking of knowledge to suit conditions.

The book, however, also looks for associative factors that count in favour of or against accepting these technologies. These factors include particular experiences (*habitus*) that affect a person's willingness to accept new knowledge. Although all knowledge derives to some extent from experience, cognition depends on mental or culturally embedded symbolic associations as well as on direct material observations. A bad experience with debt repayments, for example, may result in a farmer refusing to look seriously at innovative technologies; alternatively, the association of a technology with a politically dominant group or with an established cultural repertoire may favour its acceptance. Another associative factor in the acceptance of knowledge is its political and cultural status. Thus, the relationship between knowledge and formal systems of training, both for adults and children, also needs to be explored.

Focusing on the status of knowledge, I provide an examination of interface encounters involving resettled farmers and outside institutions and agents, since these encounters may have a bearing on how certain knowledge is perceived. Some agents might have symbolic capital (such as the backing of the state, or networks created at the local level) to make themselves heard and believed, while others might lack this symbolic capital and are not believed even if what they say is correct.

Related to this is also the examining of knowledge as a discourse and of knowledge discourses. Long (2004: 27) defines discourse as a 'set of meanings embodied in metaphors, representations, images, narratives and statements that advance a particular version of "the truth" about objects, persons, events and relations between them'. The prevailing discourses at a given time can determine what is regarded as knowledge and what is not. On the other hand when analysing discourse this book will not only consider its verbal content since

> ... discourse is never dependent solely on verbalisation of text, everyday talk or public rhetoric. It is equally manifest in non-verbal behaviour, bodily expression and feelings as well as in how people relate to specific goods, artefacts and technologies that come, as it were, already endowed with particular social meanings and valuations (Long 2001: 3).

Knowledge then can be a discourse employed by different actors to cover or hide different kinds of relationships. Thus, for example, Chapter 1 has unraveled how various socioeconomic and political constellations impact on knowledge discourses of the various actors. The chapter also discusses how the knowledge discourses of the various actors in turn shape what they regard as knowledge. It also focuses on how the various discourses impact on how the different actors involved attempt to influence the socioeconomic and political spheres.

The main *research question* that informs this book is then: How is knowledge produced, reproduced, socialised and reworked in farming areas and how do locally existing conditions filter themselves into the new practices? Hence, the book aims to accomplish three aims: (1) to analyse how social processes impact on the adoption, adaptation and dissemination of knowledge and technology; (2) to investigate how differences between actors (e.g. based on age, gender, social and economic standing, institutional affiliation, and the knowledge networks used by various actors) can impact on knowledge dissemination and appropriation; (3) to explore how existing knowledge frameworks affect knowledge analysis and acceptance and how people bridge the gap between 'outside' and 'local' forms of knowledge.

Concepts

Access to wealth, power and status

Power is defined by Weber as 'the probability that one actor within a social relationship will be in a position to carry out his own will despite resistance, regardless of the basis on which this probability rests' (Barbalet 1985: 532). On the other hand Scott regards power as the 'extent to which elites are able to impose their own image of a just social order, not simply on the behaviour of non-elites but on their consciousness as well (Mitchell 1990: 548). Weber's view of power is linked to the use of force, especially where the exercise of power is met with resistance, whilst Scott's view of power is linked to the ability and capacity of the elites to control and call upon the support of 'subordinates'.[17] Although these two views on power may be applicable under certain circumstances, they proved problematic in this study. The reason for this is that they focus more on what people can do to show power and what they cannot do to show powerlessness and fail to recognise that sometimes power is not a resource intrinsic to the elite, but instead that it can be relational. Regarding power as a relationship entails the recognition that power is not an intrinsic quality of certain beings, and guards us from regarding other people as victims of power since power is attached and detached in very different ways. As noted earlier, power is not necessarily centred in certain institutions and bodies.

All people in the sample maintained that there was no relationship between power and wealth and they denied that knowledge conferred status. They gave the impression that all people had equal opportunities of getting into positions of power in the village, and that all those occupying positions did not do so because they were rich, nor were rich because of their positions. They were also quick to point out that being a knowledgeable person or being a good farmer did not usually give that person status within the

[17] Scott's view here closely follows the Gramscian approach to hegemony that emphasises the cultural and ideological aspects of hegemony, where subject people comply with their subordination. Gramsci's theory of hegemony focuses on 'the entire practical and theoretical activities with which the ruling class (or dominant actors) not only justifies its dominance, but manages to win active consent of those over whom it rules' (Gramsci 1971, cited by Long forthcoming).

community. For example, people in one of the villages pointed out that the village headman was the poorest person in the village.[18]

Although the village headmen in the two villages have institutional power, they lack influence in the villages they are supposed to head. For example, at a meeting in Muringamombe village to discuss the policy that had been introduced by government requiring farmers to brand their cattle, most people did not pay attention to anything the village headman was saying. Every time the village headman addressed them, people would start talking amongst themselves. In Mudzinge, the village headman did not have any influence apart from leading traditional ceremonies and the people looked down on him because he was poor. The most influential people in both villages were usually the richer people or the more successful farmers and - especially just before the highly contested Presidential elections of March 2002 - those with political offices. However, it is reasonable to say that in the two villages people usually deferred to those with wealth.

Despite these denials of the link between wealth status and power, those elected to leadership positions were considered good farmers both by villagers and AREX officers. ZANU (PF) district decision-making positions, such as those of the chairperson, vice-chairperson and the youth leader, were all occupied by very good farmers who were also wealthy by village standards. Women and poor farmers occupied the less powerful positions such as those of the village police[19] and secretaries. Even when those people whom the villagers regard as 'people with knowledge' do not occupy high positions, they are still held in high esteem and they are confident and can speak out in public and make their opinions heard. Farmers who do not get high crop yields and who are generally not regarded as having knowledge do not usually speak out in public; and sometimes when they do speak, they are booed down. For example, at the Mudzinge village court, the good farmers and sometimes their wives were those who had been elected to deliberate on cases. Although anyone can contribute ideas and advice at the *dare* (village court) the people who dominate the discussions, or whose views merit serious consideration are the wealthy people regardless of whether they occupy a position or not.

People who obtained high crop yields had the opportunity not only to gain a very high status within their villages but also within their families. The reason for this was that they were in a position to assist their relatives and other villagers with the need for food or cash and other things. They managed to take part in gift giving which gave access to other resources. As

---

18 Village headmanship is a ritual office in which appointment to office follows traditional rules and procedures regardless of the wealth status of the incumbent.

19 The village police are responsible for informing people of village meetings and delivering court summons; they also maintain order at meetings and generally work as messengers.

one person pointed out, *a person with knowledge is also a person who assists his or her relatives when they are in trouble.* A person with knowledge that enables him to assist others is able to expand his patronage, and it is this patronage that gives status and influence in the villages. This could explain why some farmers fail to convert their position of relative advantage in terms of wealth and knowledge into status and power: they fail to amass a large number of clients in the village. Good villagers are supposed to share their knowledge with others; if they do not do this they will be regarded as selfish.

Despite having knowledge, a person's status in the village also depended on a variety of issues. There was a delicate balance of power between the 'good farmers' and the 'bad farmers'. To increase their influence and status, the good farmers had to know how to tread on the rope and maintain their balance. To understand this one has to understand the nature of friendships between the rich and poor. Some villagers were quite sure that apart from cheap labour the rich people did not benefit much from poor farmers. Although an outsider might regard the relationship between the poor and the prosperous farmers as basically exploitative of the poor, the relationship could be regarded as a symbiotic one. A rich farmer could get labour and support from the poor farmers, but this support could be withdrawn and given to another good farmer should the poor farmers feel short-changed in the relationship.

In some cases people considered the character of a person before voting him/her into any position of authority. A man in Mudzinge had been banned from occupying any position of authority in the village because, although he was rich and regarded as a good farmer, he had an undesirable tendency to look down upon people. However, when he was removed from his position, the position went to another rich man in the village who was considered to be of a better disposition. Although the rich usually occupied the powerful positions, they could be removed from these by local farmers, thus indicating that power was not centred in certain individuals. However, sometimes people were put into positions because they had good manners, or given jobs that no one else wanted or because they had certain qualities that were suitable for that job. For example, one poor farmer was appointed as the village policeman because he had a loud voice, which could be heard by all villagers when he was shouting to inform everyone of an impending meeting.

It would not be wrong to link being a *hurudza* (an exceptionally good farmer who is usually very wealthy due to his or her farming endeavours) and being a powerful member of society. Although people failed or simply refused to recognise this association, most of those who occupied the most powerful positions were at the same time also good farmers. Indeed, people claimed to consider qualities such as demeanour, but, except in rare cases, it would seem that such good qualities were found only among the good farmers.

If power is also defined as the ability to have influence on how other people organise their lives and the ability to instil fear into others, then other forms of power were also prevalent in the villages. If one possessed a spiritual standing in the village, as a spiritual leader, a witch or traditional healer, then one had an element of power in the village. However, different degrees of power could be accorded to these different categories of spiritual people. A spiritual leader did not have power outside his spiritual office, that of leading spiritual rituals. A suspected witch could instil fear into people but people had their own means of neutralising the power of the witch. A traditional healer gained the respect of his or her clients. Yet the power and influence born of being a good farmer was more enduring than all the other forms of power and influence as long as the power-holder knew how to use his power appropriately.

Apart from farming prowess and wealth, access to external resources can be critical for acquiring status in the village. The individual who has access to external resources can limit access to these resources to other villagers. Thus instead of the resources benefiting the community at large, they could be distributed on the basis of patron-client relationships. Although such intermediary persons might not be held in high esteem by those who received no favours, people did not want to antagonise them. Referring to the workings of one NGO, an old man pointed out that those who had received aid from the NGO had been friends of the person who represented the NGO at the village level. Thus sometimes status did not depend on how much one knew, neither was it dependent on the benevolence of other villagers; instead it depended on how one was able to network and get resources that other villagers could not.

Women usually did not attain a high status from being 'good' or 'knowledgeable' farmers, or from being wealthy. Indeed only one woman was said to be a good farmer. Although she was regarded highly by other villagers, her high esteem was not due to her farming prowess but to her traditional healing ability. Another woman was held in high esteem because they said she was a very good village health worker, although her husband was also regarded by some as a very good farmer. All the women who were held in high esteem in the village were involved with healing, either as traditional healers or as village health workers. Their being healers did not automatically confer on them a high status, but rather how they conducted themselves as healers. They had to show a concern when people approached them for help.

Wealth and poverty

At the time of resettlement no poor households had any cattle, while only two of the households in the medium-wealth category did not possess any.

Those who had a few resources (such as cattle or ploughs) to their name managed to get a headstart over other villagers who had to depend on manual labour or borrow cattle from others. Ownership of cattle alone, however, can not wholly explain the social differentiation that has occurred in the two villages since the 1980s. This can be understood when we take into cognisance that, at resettlement, people could receive two oxen on loan from the Agricultural Finance Corporation. Some of the households in the sample, including one household in the very poor category, had access to this facility. What is important at this point is to understand why some people who were poor (i.e. did not own any cattle or ploughs) at the time of resettlement managed to improve their condition, while others remained poor.[20]

In their concern for how definitions of poverty affect the results of the extent of poverty in eight industrialised countries, Hagenaars and Vos (1988: 212-3) advance three categories or definitions of poverty:

(A) Poverty is having less than an objectively defined absolute minimum (absolute poverty);
(B) Poverty is having less than others in society (relative poverty);
(C) Poverty is feeling you do not have enough to get along (subjective poverty).

For the ranking scale we adopted the absolute poverty approach. As will soon become apparent, category A and category B tallied with farmers' perceptions of poverty since for them poverty was relative and subjective. However, in order to measure wealth and poverty it was not proper to rely only on farmers' perceptions since farmers who apparently suffered extreme deprivation did not regard themselves as poor even though sometimes they had to work for food in the fields of better-off farmers. On the other hand, those farmers who had access to good food, cash and better lives than others sometimes did regard themselves as poor for some other reasons such as a lack of relatives (discussed later in this chapter in the section on the concept of a good farmer). Thus there was a need for other objectively identifiable indicators based on local farmers' answers to certain questions. Farmers were asked to identify what made a person a good farmer, what distinguished a good farmer from other farmers, who in the village they thought were rich and who were poor and why; and what kinds of lifestyle in the villages were associated with poverty and which were associated with wealth. It emerged during these discussions that, although farmers were

[20] Note that, although some households have remained poor according to the arbitrary scale of wealth ranking we applied, this does not mean however that there has been no marked improvement in the life of these people. It also should be noted that some of the medium wealthy households might be slipping into the poor category.

sometimes not willing to refer to themselves as poor or rich, during a general discussion on wealth and poverty certain recurrent themes emerged which then allowed us to rank households within the sample according to wealth.

Taking into consideration farmers' perceptions of poverty and wealth, in order to assess the wealth of households, we developed an index as follows:

Table 2.1 Calculation of wealth

Cattle ownership	0=0; 1-2=1; 3-5=2; 5-9=3; 10+=4	0-4 points
Ownership of equipment	Ploughs, scotch carts, cultivator, harrow, tractor	0-4 points
House	Cement, plastered floor, and windows of main house, roof of extension	0-4 points
Other investments	Grinding mill, rental house in town, solar panel, other significant livestock, etc.	0-4 points
Maximum range		0-16 points

Using this arbitrary wealth ranking scale, it emerged that in the sample, 2 households were very poor (0-1), 5 were poor (2-6), and 7 were medium (7-11), whereas none were in the very wealthy category.

Elsewhere Poulton *et al.* (2000: 5) have characterised poor households as follows:

> Poor households tend to be characterised by the following features: large household size, high dependency ratio, older or very young household heads, small land holdings and low levels of education ... Poor households tend to be food crop farmers ... Livestock holdings are a key indicator of wealth (and a critical production asset) amongst small holder households.

However, as will soon become apparent, although some households were large and poor, and others were large and rich, and sometimes smaller households could be very poor and depended on food purchases, Poulton's *et al.* (2000) characterisation of poor households generally holds true for the poor households in the study sample.

Although one of the households that improved from the very poor category to the medium-wealth category consisted of sixteen people, eleven of whom provided labour, another of the very poor households consisted of only one individual who received no outside help. As a result, although the availability of labour seems to be an obvious variable, we are loath to attribute all the wealth differences to the availability of labour. One of the reasons for this is that at resettlement most of the families were still very young with young children who could not be expected to do a day's work. On the other hand, at the time the fieldwork for this book was carried out, one of the poor households had fourteen members, all of whom provided labour, whilst one of the medium-rank households had three members who also provided all the labour requirements and sometimes hired one or two people to assist.

One variable that clearly distinguished the poor and very poor households from the medium-wealth households was that the former did not have regular access to seed and fertiliser loans. When they could get these loans, they usually obtained very few inputs since at times the amount of a farmer's loan depended on his/her previous harvest and the quantity of crops the farmer had delivered to the loan bodies for sale in the previous year. As a result, in most cases, these poor households had poor yields since they resorted to the use of saved seed and little or no fertilisers. Access to loan facilities acquires a larger meaning if it is sufficiently realised that much farming in these villages was dependent on access to these loans. On being asked why they were very poor when her husband was still alive, one woman stated that,

> This is so because we are now getting more fertiliser on loan from the GMB. Without fertiliser, there is not much anyone can do. Before my husband died he bought 3 cattle that we use for ploughing now we have got draught power. When my husband was alive, we could not get credit from GMB because we did not have cattle to use as collateral. When borrowing fertilisers and seed packs from the GMB they want to know if you have cattle, scotch carts, ploughs, and any other farming equipment.

Whilst most of the poor households were denied loans because they were bad debtors, two of the medium-wealth households sometimes chose not to take out any loans because they could obtain all the required inputs on their own. A third household had not taken out seed and fertiliser loans at one time due to AIDS and HIV-related problems.

Poverty levels might also have something to do with people deciding to join the resettlement schemes. Most of the better-off households defined their core business in the resettlement scheme as farming. They had migrated from their original home areas because they were very much interested in farming and therefore in these areas with good soils and rainfall. On the other hand, three of the very poor people had migrated from their areas as a way of avoiding conflicts and running away from witchcraft, which they said was being practised on them by their relatives. These people were not much concerned with farming and were mostly satisfied with their fate saying that in the resettlement areas they now enjoyed peace of mind. As long as they obtained enough to feed their families, they were not concerned with many things.

When asked to describe a 'good farmer' most respondents, regardless of their wealth status, said that good farmers were good because they planned their things on time, while on the other hand, those who were not good farmers began to run around trying to secure seed after the rains had started. Even if he got seed and fertiliser loans on time, one of the very poor household heads in the sample sold the fertilisers and seed very cheaply to other farmers in the area. He would even slaughter the cattle he had received on loan and sell the beef. Even though at one time this man had two wives and six children who could provide the required labour, his household was always short of food and had to purchase supplementary food. Also, like in one of the polygynous households in the sample, the wives would withdraw their labour if they thought the other co-wife was foot-dragging.

Judging by their decreasing herd of cattle and quality of life, one household noted that they were actually worse off than they were prior to resettlement. The wife of the respondent explained that this was because they had been bewitched by their husband's relatives just before resettlement, something which had also influenced their decision to apply for resettlement. The difference between this household and the other households in the medium category was that the household head was of an advanced age (80 years old). He was different from other households in that his field had been broken up into plots to accommodate his married sons and unmarried daughters with families. The fragmentation had meant that the area of land available to him for food and cash-crop production had been drastically reduced and that he was instead mostly concentrating on food production. The fragmentation had come at a time when he was saddled with the burden of looking after his very young grandchildren after the death of his two children from AIDS. Worse still he had not managed to farm properly and to secure any loans when his two daughters fell sick. His daughters' illness had reduced the available labour since they needed constant attention. Thus, although there may be other factors to account for why some families are regressing and fragmenting their fields, such as the stage reached in the household cycle, illnesses within a family can reverse the fortunes of people.

Access to enough land might also be a necessary factor in explaining poverty (see the table below, which shows the number of acres the different households own by category of wealth).

Table 2.2 Land available to various social categories (in acres)

Acres available:	0-5	6-9	10-12	13-25
Very poor	1	1	1	-
Poor	-	2	2	-
Medium	-	1	2	4

More than half of the successful farmers managed to expand their fields, while some of the poor farmers have diminished their acres. Sometimes, although the poor have access to fewer acres, they are not able to cultivate all their land due to a lack of labour and other resources. Sometimes poor farmers with an average 12-acre field will rent out part of their field to more successful farmers, or to teachers and other business people, in return for bags of fertiliser to use in the part of the field they are utilising.[21]

Gender

Waterhouse and Vijfhuizen (2001: 7-8) maintain that gender concerns the social-cultural construction of women and men. According to them,

> ... it is necessary to realize that these processes of construction are both reproduced and transformed, by both women and men. Hence analysing gender relations means getting to grips with these processes. It must be emphasised that both men and women are involved in constructing gender, by their actions (practices), assigning meanings and reproducing norms and values. From this perspective women are seen as strategic actors and not passive victims of patriarchal and matriarchal structures.

This section will briefly look at the social construction and reconstruction of gender roles by men and women in the resettlement villages I studied. In these villages, women were not resettled in their own right (Jacobs 1990; Gaidzanwa 1995; Rukuni 1994). This is not an experience unique to Zimbabwe but, as noted by ILO (1995: 26, 27),

> ... reforms have almost always defined beneficiaries of land titles and of any concomitant support services to be the male head of the household. This proved to be disastrous for women who had enjoyed customary use rights and for female heads of households ... Women's access to agricultural land in resettlement areas ... has not always been secured. First, schemes often grant land titles or leases to male heads of household. Second, where land acquisition or allocation is open to everyone, men or women, a number of socio-cultural and economic factors prevent women from enjoying or exercising this equal right.[22]

---

21 In her study of resettlement areas Jacobs (1990: 173) found similar trends. She noted that those farmers who lacked draught power or equipment with which to cultivate, often rented out land and thus in her study 5 per cent of resettled farmers were utilizing between 14-37 acres of land instead of the designated 12 acres only.

22 Commenting on land tenure commissions (which were led by Rukuni) in Zimbabwe, Mbiba (1999: 316, 322) points out that 'they marginalise the majority of peasants and exclude any radical submissions from the affected people as became clear in the handling of gender issues by Zimbabwe's Land tenure Commission (1993-1994). Women comprehensively called for gender

In Mudzinge, one woman managed to obtain a plot permit on her own when her husband, who had been the initial receiver of the land, ran away with another woman and did not turn up to claim the land. Other women in the sample became plot owners and *de facto* household heads upon the demise of their husbands. In Muringamombe for example, both women in the sample became heads after the death of their husbands. Considering the large number of women in the area, particularly taking into account that some men have more than one wife, there is a gender imbalance in the ownership of stands.

Even in those households where the woman is the head, she remains household head only in as far as her name appears in the official documents. Most of the decisions on farming are taken out of the hands of women. Thus, simply counting the number of female household heads versus the number of male household heads tells us little about gender politics in the area. Where there are older sons, the mother, who is the official household head, defers most of the important decisions to the grown-up sons. These decisions may be related to land use or to the commercial crop varieties the family ends up cultivating.

The household structure is such that men make most of the decisions, even those related to the disposition of produce and how the proceeds are to be distributed within the household. This unequal distribution of power in the domestic sphere might also be the cause of the many domestic disputes which frequently call for the intervention of the village *dare*: women are contesting inequality in their homes because most agriculturally demanding tasks are undertaken by women.[23] These gender inequalities are not only limited to the domestic sphere but extend to public power domains as well. Men occupy most public positions that are regarded as powerful.

Despite these gender differentials, it seems that both men and women have access to the same kinds of information. The differences that arise concern how this information is processed and used. Some women do not take loans or, if they do, they only take small loans because of the belief that taking loans is men's business (*zvinotoda varume*). Moreover, the socialised

---

equity in land resource use through allocation of land to women ... Yet in the commission's 300 pages report evidence from women on this score was ignored in preference to maintenance of the male-Chief *status quo*'.

23 Note that in her study of the Nso in the Cameroon Grassfields, Goheen (1996: 73) highlights the importance of women in performing agricultural tasks: 'While doubting women's capacity to reason, men continue to acknowledge women's farm labour as the most critical factor of production. Women are believed to be naturally endowed with the capacity to do farm work. When questioned about food crops or indeed about farming in general, men would often say "you will have to ask the women. It is they who are being the farmers, and it is they who know about farming" ... over 90 per cent of the food is grown by women'.

belief that men are more capable than women has resulted in women not having confidence in their own capacities.

Men are also able to command much labour. For example, there is a general sentiment among male stand-holders that it is the women who are the labourers. Some villagers equated having many wives with the ability to farm their whole 12 acres of land, or at least most of it.[24] In her study of resettlement areas in Zimbabwe, Jacobs (1990: 173) notes that 'some men may hope to pursue polygyny, as a strategy for accumulating capital through accumulation of wives ... small capitalist farmers who were polygynists treated wives (especially younger ones) virtually as labourers'. Similarly, in the two villages I studied, a man was usually said to have married the right woman if the woman was able to work very hard. Marriage for any other reason was viewed to be an error of judgement. I talked to one female trader who remarked that men oppress women since they do not allow their women to use the maize to get what they want, despite the fact that women do most of the fieldwork. The ability of men to marry labour gave them an edge over women whereas, on the other hand, women depended on their children's labour, which they could lose at any time if their children married and moved out.

Household conflicts over resource-sharing indicate that proceeds from farming are not distributed equitably within the household. Thus, the same resource can impact differently on different people within the household. In one of the households in the sample, the wife had to take her husband to the village court because she felt that the proceeds from farming were not being distributed fairly. According to her, the husband was always buying cattle, which, culturally, is not regarded as a woman's property and could be taken over by her husband's relatives upon his death. This demonstrates that resources distributed by the state (in this case land) ostensibly to improve people's livelihoods sometimes do not have the same impact on different members of the targeted households. Discussing development and state policies from a gender perspective, Parpart and Staudt (1990: 1) express misgivings that

> ... women's seemingly personal, everyday experiences are structured by policies, most of which are apparently 'gender neutral'. But these policies are, in fact, experienced very differently by men and women. We have been struck by the absence of attention to women and the state in several different bodies of knowledge.

---

[24] Cheater (1984) found a similar kind of perception among Purchase Area farmers in Msengezi which she calls 'marrying labour', when men married several wives so that they could get cheap labour to work their farms. In another study, Weinrich (1983) documented that 47 per cent of the families in Purchase Areas in southern Zimbabwe were polygynous.

In another case, a man who was married to a female stand-holder (not in the sample) took his wife to court because he felt that she was controlling all the income from their farming ventures. The wife, on the other hand, argued that she was investing the money in family assets and that her husband was mentally disturbed and could not be trusted with any household money. There were also cases where husbands beat their wives for exchanging all the maize for clothes and other utensils with the traders popularly known as *madhaiza*, who exchange their wares for maize and sometimes cash. Men deplore this practice, as they say it threatens the family's food security, and at times men have chased these *madhaiza* traders away from their homes. Some women, on the other hand, see this as fair practice since men use all the money from their cash crops to buy manly things or to purchase goods that the women cannot call theirs. So, buying from *madhaiza*, women could purchase items for the benefits of the whole family and, on top of that, utensils that they could claim to own.

Thus, even in cases where there is an observable increase in the asset base of individual families, this does not necessarily lead to the improvement of the livelihoods of everyone in the family. Although Kinsey (1999) points to the positive effects of resettlement shown by certain economic indicators on household poverty levels, he fails to notice that these perceived increases in household wealth could be experienced quite differently by different people.[25] Thus sometimes, although a household can be positively ranked in terms of available wealth, it does not necessarily follow that the members of that household enjoy the fruits of that wealth in equal measure. A focus on gender thus enables us to keep in mind that people within the same household often experience phenomena differently.

Sometimes conflicts would also occur between parents and their adult children. In some cases, adult children ended up beating or threatening to beat up their parents because of frustration over the way resources were being shared in the household. Even in those households that were regarded as well-off and whose household head was seen as someone who invested wisely, these conflicts could occur.

The concept of 'good farmer'

For officials the concept of 'good farmer' was associated with people who achieved higher yields than most people did in the village, yet farmers talked of *hurudza*. Whereas the concept of 'good farmer' registered success in agricultural activities, the concept of *hurudza* was associated with lifestyle.

---

[25] This brings to fore Sen's (1981: 1) disenchantment with the focus on food availability, because for him it blurs the important issue of access and control over food. According to him, 'scarcity is the characteristic of people not having enough ... it is not a characteristic of there not being enough ... While the latter can be the cause of the former, it is one of the many causes'.

Hurudzas were taken to be those farmers possessing status goods such as solar panels, television sets, radios and other household items like lounge suites. Consequently, for local farmers a person could be regarded as a good farmer but still not be a *hurudza*. Those who were regarded as good farmers by the 'experts' and not as *hurudzas* by local farmers were usually suspected of practising magic. This was so because villagers believed that money from magic could not buy anything. Everyone agreed that a *hurudza* should have status goods such as solar panels, televisions and sometimes a car and production goods such as cattle, as well as the ability to hire labour. And some people went even further to claim that a *hurudza* also had to have good clothes and good food which was identified by the variety of food stuff that he could obtain. Therefore the concept of *hurudza* is not only associated with the ability to achieve high yields but also with decisions taken on disposable income. Good farmers were taken to be those with adequate draught power, implements and the ability to employ other people.

Farmers who had big yields, but used loans, were not considered good farmers. The general belief was that good farmers should have their own independent sources of income. They were supposed to finance their own agricultural activities rather than work as contract farmers for credit organisations.

Furthermore the concepts of 'good farmer' and 'poor farmer' were not seen as exact opposites. Although three farmers thought that the concept of poor farmer could be used interchangeably with the word laziness, the remainder of the people in the sample felt that poverty was not always due to this. If some people could work well when working for money in other people's fields., it was felt that lack of access to relevant resources could be the only explanation for poor performance in their own fields. Just as the good farmer was associated with access to adequate resources, so the poor farmer was associated with lack of access. The major resource that was mentioned, regardless of gender or economic status, was access to cattle. Lack of cattle was felt to be the major cause of poverty since it meant that a person always planted crops late and consequently received poor harvests. Poor yields reinforced poverty within the household since it implied a lack of food and cash to buy agricultural inputs or other household essentials.

Three people in the total sample pointed out that if a person was disabled or did not have any relatives then that person would be poor regardless of the amount of crops he harvested. Consequently, poverty was not always related to farming ability. And, on the other hand, some people categorically denied that they were poor, even though they fell into the 'poor' category according to our wealth-ranking criteria.

Knowledge as a concept and as a practice

The following four illustrations provide an exegesis on what villagers thought about knowledge as a concept and a practice:

Case 1

This example is about a farmer who the villagers generally agreed was very lazy. He was an alcoholic and usually obtained no more than a wheelbarrow of maize from his twelve-acre field. Furthermore he had chased away his wife and children and he lived alone. Regardless of this, people maintained that he had a sharp mind and was very knowledgeable. He was a very exceptional builder. He usually gave people very sound agricultural advice which, if they followed it, would sometimes work very well. One woman pointed out that he was poor not because he did not have knowledge but because he was too lazy to work. When it came to electing people to village positions that had nothing to do with farming, they usually elected him because he was a good public speaker and they knew he would represent them well and would not embarrass them in relation to outsiders. However, they could not elect him into positions related to farming because he was always away and did not have any interest in farming.

Case 2

This man was from another village but he was very much like the man described above. He presently had no wife and no children and people had lost count of the number of wives who had jilted him, though this number was said to be no less than seven. People said that the wives ran away with other men because the man was lazy and could not feed them. The general consensus was that women like to stay where they are properly fed and well looked after. This man was a woodcarver of renown, his hallmark being that even white men bought his work. People in the village agreed that this man had great energy and they usually invited him to work for money in their fields as he usually did a perfect job.

> It is not that he is not strong and healthy; he just wants to work for other people for money. When you ask him to work for you for such and such an amount that is when his farming spirits (*shavi rekurima*) possess him, but not in his field. He does not even see that if he harvests a good crop he will be able to buy his own things. We can say people like that do not know farming regardless of how good they are when they work in other people's fields.

Case 3

In one of the villages, there was a Master Farmer. There was a consensus between the AREX officer and the villagers that this farmer was very good, but most villagers, including six out of the seven in the sample, maintained that his being good did not mean that he had knowledge. He was not at all popular with the villagers because he was arrogant. He was not regarded ignorant because he did not invest his money for the benefit of his family.

Most villagers pointed out that he did not have good matrimonial relations with his numerous wives as he was known to beat them sometimes and he was not on good terms with some of his in-laws who lived in the village. He usually did not buy anything for his wives but instead bought things his wives could not use such as the motorbike which he had recently acquired. Instead of buying a car which would be beneficial to the whole family he used the money for himself and spent it on acquiring other women and girlfriends when in fact his wives did most of the work.

When I pointed out that this man should have knowledge because I had heard from other villagers that they usually held field days at his field, one old man admitted that field days were usually held Magambas' field or Samson's.

> However, now because of Samson's bad manners, no one wants to go to his place for field days. If we hold a field day at his place, afterwards he struts around the village telling people that he is the only good farmer around. Instead of the field day being a learning experience, it becomes a shaming experience. On the other hand, the AGRITEX Officer we used to have went away and we were left with an officer who was cruel and did not want us to succeed. Mrs R. used to encourage the AGRITEX Officers to always visit the farmers. She was a hard worker that one.

Case 4

Another good farmer in the sample had a good relationship with the AGRITEX and was popular with fellow villagers. Villagers stated that he was good but did not have any knowledge because his children were badly dressed and did not eat good food.

These four cases help us understand what the villagers regarded as knowledge. While 'experts' focus only on agricultural knowledge, farmers are much more concerned with more general social knowledge. The first farmer was lazy, but he was seen to have knowledge because he was a good public speaker and an exceptional builder. He also sometimes offered people very good advice. Although the second case was somewhat similar to the first, and although this man could perform agricultural tasks successfully in other people's fields, he was still not regarded as a person with knowledge. The third one was a very good farmer but was not considered knowledgeable because of the poor relations he enjoyed with other villagers as well as with his numerous wives. The fourth farmer was also a good farmer but not necessarily one with knowledge because his family did not live in a style a farmer of his calibre would deserve.

Hence, farmers use double standards in judging whether a farmer has knowledge or not, so that these judgements are more an indicator of social relationships than of anything else. Villagers judged farmers on the basis of their overall lifestyle. The concepts of knowledge and 'knowledgeable persons' varied over time, and were not wholly reliant on farming ability. Experts focus mainly on whether the farmer is 'good' or 'bad' at farming and whether he is adopting their advice. They are not concerned with how the farmer relates to other farmers in the area. This is evidenced by the insistence of AGRITEX to hold a field day at one of the good farmers' fields, ignorant of the fact that people did not him like because of his arrogance.

Still, with regard to knowledge, some respondents believed that everyone had knowledge while others insisted that some had more than others.[26] As a result, farmers accepted that everyone had knowledge but evidently some forms of knowledge served the farmer better, since he achieved crop yields.

Thus although experts envisaged a positive association between good yields and knowledge, for the resettled farmers having farming knowledge did not necessarily entail good yields. To some extent, good crop yields were seen as dependent on people's relationship with their ancestors. If the ancestors were well pleased with a person's conduct they could richly bless him/her, but if they were not, they could show their wrath by giving the person poor yields regardless of how much s/he knew or of how many resources s/he had. Secondly, while knowledge experts (e.g. AGRITEX, veterinary officers and loan organisations) considered knowledge in a highly individualised and economised context, villagers interpreted the concept of knowledge in a broader social context. For the knowledge experts, proper tutoring could move a person from a position of 'ignorance' to that of 'knowledge', whilst for villagers a person might gain more knowledge, but was never ignorant in the first place.

Farmers with knowledge[27]

There were, however, at least three farmers who regarded other resettled farmers as lacking in knowledge. These farmers regarded themselves as good, knowledgeable farmers. They usually took pride in their association with outside agencies such as AGRITEX, Zimbabwe Tobacco Association, Agribank, or Zimbabwe Farmers Union, or in their previous interaction with

---

[26] As will be explored in a later chapter, the notion was different when it came down to women. Some respondents said that some women did not have knowledge at all, whilst others acknowledged that there were some women who had more knowledge than some of the men.

[27] It is interesting that these three farmers who regarded themselves as farmers with knowledge were not regarded this way by fellow villagers, although all of them were seen as very good farmers.

white commercial farmers. These farmers did not have much in common except that two had worked in former commercial farms where they said they had acquired their knowledge, arguing that commercial farmers know even more than the AGRITEX officers do.

One of the farmers was relatively highly educated and claimed to have finished standard six. He also attended agricultural courses wherever they were held and, although he had never worked in the commercial farms, he felt that if he secured enough resources he could perform just as well as the commercial farmers. Association with outside agencies usually meant that some people could obtain certain information not available to others. For example, through associating with the Zimbabwe Tobacco Association, one farmer was able to receive free training, which he later applied in his farming activities. However, other farmers who were not resistant to acquiring knowledge from outside preferred the knowledge to be practical rather than theoretical, and were therefore not very keen on attending theoretical courses. Farmers were more likely to adopt certain practices they had witness to be successful elsewhere than adopt things that they were told worked in a class-room situation.

Surprisingly, however, this distrust of theoretical knowledge did not extend to school knowledge. This also counters what Bourdillon *et al.* (2002) found in Wedza district where knowledge from school was referred to as 'book knowledge' that lacked practical relevance. Everyone in the sample agreed that school knowledge was important to agricultural performance even though some of the respondents did not have children who had gone to school or who were attending school at the time. However, when asked if possession of school knowledge made a person a good farmer, one farmers who believed he was good and had knowledge remarked that theoretical knowledge meant nothing until the person who had it could apply it successfully. Unfortunately this was not always the case. Thus, for farmers, the celebratory claim by the officials that they had managed to disseminate knowledge to a great number of people through the Master Farmer programme did not mean anything if that knowledge could not be translated into observable production prowess.

These farmers, who regarded themselves as more knowledgeable than others, usually spoke differently and had a different understanding of farming. They were against most traditional practices such as the observation of *Chisi* (rest days), which, they said were time-consuming. These farmers were more likely to use English technical terms such as 'metres', 'centimetres', 'kilograms', 'acres', 'plant population' and 'energy'. The other farmers used terms such as *madhunduru* (contour ridges) or *madrain* (water channels), to refer to areas they had under cultivation, or referred to 'bags' of fertiliser instead of 'kilograms' when talking about the fertiliser they had used in their fields or 'scotch carts' instead of 'tons' when discussing yields.

These farmers 'with knowledge' usually looked down upon others whom they thought had 'no knowledge' or at least upon those who shunned knowledge agencies. For these people, one of the reasons why some people performed badly was that they did not have knowledge.

Same thing, different terminology?

'Experts' and the common man encounter the same issues that they have to deal with. Even when they apply the same method to discuss certain issues or use a particular terminology, do they understand the same issue in the same way? Understanding terminology is a very important part of the dissemination of knowledge, to ensure that all parties involved are talking about the same thing. A good example is the concept of crop breeding. AREX officials noticed that farmers who planted well-bred seed from the seed companies and applied the required fertiliser did very well, as compared to those who planted saved seed. Most farmers agreed that saved seed, particularly the new Seed Co varieties, were not very good. Saved seed, with the exception of some Open-Pollinated Varieties (OPVs), was never ranked positively. Farmers instead preferred *mbeu dzakauchikwa*. *Kuuchikwa* is a traditional process usually associated with helping infertile couples to become fertile. AREX adopted this term when explaining the new hybrid varieties to people.

However, evidence indicates that the way the two parties understand this concept of breeding/ *kuuchika* is different. For AREX *kuuchika* is the process that takes place in scientific laboratories where new breeds of seed are developed and seed manufactured. The process of *kuuchika* ends at the factory gate. Farmers only need to follow the requirements of the technological packs to obtain good crops. On the other hand, for farmers the process of making the seed fertile does not end at the gates of the Seed Company. In some cases it does not start with the company but with the farmer him/herself. The headman of Mudzinge village maintained that for crops to do very well people had to take their seed to the *zumba* (traditional hut) for them to be blessed before they were planted. At the *zumba* the seed would be made more fertile by the power of the spirits. Some of the villagers left their seed in the kitchen overnight, at the *huva* where the spirits would supposedly bless them to ensure good harvests.[28] Even those who did not believe in the traditional method of blessing their seed used other ways. Others asked for holy water from priests to sprinkle into their fields to chase away evil spirits and ensure healthy crops.

This makes it clear that at times experts and local people can use the same terminology but refer to qualitatively different things. To understand the social processes surrounding the production and dissemination of know-

[28] The concept of *Zumba* and *huva* will be discussed in detail in chapter 7.

ledge, the following chapters will pay attention to the different meanings assigned to processes and phenomena.

From knowledge to specialised ignorance

Based on the available evidence, it is possible that, as people acquire more knowledge, they move from a position of knowledge to one of specialised ignorance. By 'specialised ignorance' people become more competent in doing certain things related to a certain task, but at the same time increasingly acquire ignorance in doing certain things associated with the complete task. For instance, the first generation of settlers in the resettlement could save their own seed. They were not overly dependent on seed companies for seed and could preserve their own consumption maize using other means but not modern chemicals. However, people become more ignorant in these spheres as they acquire the skills and status of the modern farmer. Traditional ways of doing things become down-graded and associated with the more 'ignorant' or 'poor' farmers. As farmers become modern and more knowledgeable they become more dependent on seed houses and agro-industry for their needs. Those who still use other methods of farming or who keep securing their own needs are regarded as ignorant and backward since they have refused to move with the new technological trends.

There is another interesting twist to the conceptualisation of knowledge by the resettled farmers. Knowledge is usually associated with practice but, as people obtain more 'expert knowledge', the concept becomes diluted. People can lay a claim to knowledge not because they are seen to be able to do certain things but because they have been taught or told how to do those things. One of the farmers, who said that one should consider crop yield when identifying knowledgeable farmers, also claimed that his children knew how to preserve saved seed and maize for consumption. Although his children had never performed these tasks because they were dependent on commercial products, they knew how they should do them because he had taught them, in case they ran out of money to buy the requisite seed and chemicals after he was dead. Thus, the beginning of knowledge could at the same time imply the beginning of 'the growth of ignorance', to borrow a phrase from Hobart (1993). According to Hobart the growth of knowledge and ignorance are linked because as experts acquire more knowledge they designate other modes as ignorant. In this Zimbabwean case, even though the experts still label many traditional practices as 'ignorant', the more knowledge 'modern' farmers acquire, the narrower their knowledge vision becomes and hence the more ignorant they are of alternatives.

Conclusion

This chapter has positioned the study in relation to various theoretical considerations and has argued why knowledge has to be studied and understood in context. There was also a concerted effort to identify concepts such as wealth, poverty, power and 'good farmers' as they are defined by the various actors (i.e. experts and farmers) involved.

I also explored why knowledge experts and farmers tend to approach phenomena differently, giving rise to different perspectives on knowledge. For example, we find important differences in the definitions that 'experts' and 'lay farmers' gave of the notion of 'good farmers'. Moreover, sometimes these discrepancies led to situations wherein farmers expressed disappointment with the pronouncements made by experts, not because the latter were intrinsically wrong but because they violated certain strongly held farmer assumptions.

This chapter has also made it abundantly clear that programmes and policies are not experienced in the same way by different people. For instance, policies that claimed to be gender-neutral were in fact experienced and perceived differently by men and women. This highlighted the central dimension of this book, namely the need to consider households and local farming categories as being composed of actors who have both common and divergent interests.

Finally knowledge is not always what it seems and is not always a positive thing. Knowledge can even be disempowering to those who are equipped with it. Although modern scientific knowledge is efficient, it has made farmers more dependent on agro-business as opposed to the more autonomous space they enjoyed when all their sources were locally available.

3

The research context

Photo 3.1 Behind the two women is Mr Karidza's gota where maize is kept to dry before shelling. Photo by author.

Introduction

This chapter serves the purpose of situating the reader in the research setting. It provides a background of the history of farming in the two study villages, a picture of the agricultural servicing institutions in the area, and a brief discussion of the significance of kinship and religious affiliation.

Photo 3.2 Delivering cotton bales to the Cotton Company's depot. Photo by author.

The history of farming in Mudzinge and Muringamombe

In the first year of resettlement, the government cleared and ploughed an acre of land free of charge for each farmer. Farmers also received maize seed and fertilisers for their ploughed acre. In offering assistance, the government emphasised maize as the food crop *par excellence*?. However, people also received groundnut seed. This history of settlement differs slightly from household to household depending on when they joined the scheme. Those who arrived a year or two after resettlement received none of this free aid.

Even in the early years people planted some other crops such as groundnuts, cow peas, beans, sweet potatoes and soya beans. However, AGRITEX focused mainly on advising people on cotton and sometimes maize. This emphasis has not changed, although recently there has been an observable shift to tobacco.

People focused mainly on maize to achieve household food self-sufficiency and also, as one household head in Muringamombe pointed out, 'to feed the nation'. On the other hand cotton was stressed as a cash crop to enable people to participate in the cash economy. As a result, throughout the history of resettlement, maize has been cultivated primarily as a food crop, though any surplus could be sold for cash to the GMB. One respondent commented that

> ... when we plant our maize we are very happy when we sell it to the GMB because we know we will be feeding a lot of people. Even if there is a drought we know that our national granaries will be full, so that we will not have to beg from other countries ... Those are the important aims we consider when planting maize. Therefore, we don't plant maize only for the consumption of the family but for selling. If I get money after selling the maize, that's good on my side, selling maize is also good for the other person because he will have access to food.

Thus in Mupfurudzi maize, tobacco and cotton can be regarded as the major crops. People still cultivate crops such as groundnuts, cow-peas, sweet potatoes, round nuts and beans but these are regarded as lesser crops or women's crops that are cultivated to augment the family's food needs. However, it has to be noted that the domestic market for these crops is flourishing and pays well, sometimes far above the official market prices. Women also control the domestic marketing of these crops. The history of marketing in this area indicates a shift from subsistence to commercial crop farming.

As shown in Table 3.1 and Figure 3.1 below, with regard to maize there has also been a shift from the old High Yielding Varieties (HYVs) of maize to second generation HYV maize. However, what the table and figure do not highlight is that people still depend on Open-Pollinated Varieties of maize such as Hickory King and additionally make use of saved seed.

Table 3.1 Percentage of land sown with new varieties by type of new variety by year, Mupfurudzi

Year	SC40x (%)	SC50x (%)	Sc60x (%)
1994/95	4.0	94.0	2.0
1995/96	11.9	83.6	4.5
1996/97	18.8	73.8	7.5
1997/98	20.8	69.5	9.7
1998/99	28.5	57.1	14.4
1999/00	26.8	55.4	17.8
1998/99	28.5	57.1	14.4
1999/00	26.8	55.4	17.8

Source: Bourdillon *et al.* 2002

Thus, when asked about the variety of seed they cultivated, even those people who planted saved seed would mention the original variety, thus giving a false impression. This is because people identified the saved seed with the original seed variety.

Figure 3.1 Acreage planted to maize, Mupfurudzi

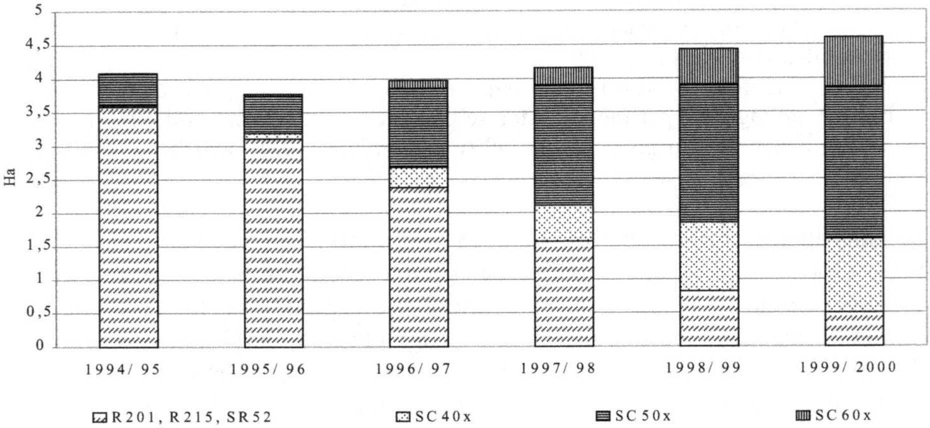

Source: Adapted from Bourdillon *et al.* 2002

For Figure 3.1 above, the numbers R201, R215, and SR52 refer to maize varieties. SC40x, SC50x and SC60x indicate the series of the maize variety. For instance, for the SC40x series there are many varieties that include the SC401, SC403, SC405 and SC407. Figure 3.1 and Table 3.1, both adapted from Bourdillon *et al.* (2002), indicate that people have adopted the new varieties of maize in a wholesale manner, and that the portion devoted to the old varieties has decreased. The shift to these new High Yielding Varieties of maize is mostly due to the fact that the old varieties that people prefer can no longer be found in the shops (see Bourdillon *et al.* 2002).

A brief background of sample households

The resettlement scheme was composed of a medley of individuals from different walks of life and different parts of the country. In the sample of fourteen households, four households were resettled from the Large Scale White Commercial Farms. Three of the household heads had worked and lived with their families at Farms (LSWCF) in Mashonaland Central Province. Two of the three household heads were of Zimbabwean origin but had decided to migrate from their rural homes to the farms in search of work, and to escape social conflicts in their rural homes, which included witchcraft accusations and suspicions that their relatives where performing witchcraft acts on them. One was from Maramba (in Murehwa Moshonaland East) and the other one came from Madziwa Communal areas near Mupfurudzi resettlement scheme in Mashonaland Central. The third household head was of

Mozambican origin and did not have a rural home in Zimbabwe. The fourth household head was resettled from a LSWCF in Karoi (Mashonaland West Province), was of Zambian origin and did not have a rural home in Zimbabwe.

Four households in the sample were female-headed. As noted earlier three of these women became female plot holders after the death of their husbands, whilst the fourth became a plot holder after the husband who had been the initial recipient of the land had run away with another woman. The first woman came from Rusape (Manicaland province), the second from Chiweshe (Mashonaland Central), and the other two came from Murehwa in Mashonaland East.

Five household heads had been resettled because they were regarded as landless peasants and urban refugees. Although one of the household heads had a rural home in nearby Madziwa communal area, he only had a piece of land to build a home to accommodate his family and had none left to farm on. As a result he had migrated to Harare to seek employment. At independence he applied for resettlement and when he was resettled he left his job. The second household head's situation was more or less the same, with the exception that although he was based in Madziwa Communal area and had no land he did not have an urban job. The third household head was form Mutare in Manicaland Province. He worked in Mutare town but he had no access to land. He had bought some cattle which he had to ask his relatives to look after since he had no access to land. At independence he also applied for resettlement and left his work. The fourth and fifth households were resettled from Chiweshe. Although they were not formally employed they were both builders by profession so they made money by building houses for people. However, they did not have access to land in Chiweshe as a resulted opted for resettlement.

The fourteenth household head had been resettled from Chimanimani in Manicaland province because of his disability. He used to work as a driver for a construction company in Harare. However, he had lost a leg in a road accident whilst on duty. He strongly suspected witchcraft to be the cause of that accident. He was on disability pension and also owned a house in Harare.

Because there were some people within the resettlement scheme who came from the nearby communal areas, cordial relationships existed between the resettled people and the local people. When it was time to choose a village head, one of the resettled people who came from a nearby communal area was chosen as the village head, as he had the same totem as the royal lineage and also because being a local it was expected that he understood local cultures and customs. Strong ties also exist as a result of intermarriages that exist between the resettled people and the local people. At the same time totems can be used to create fictive kinship. For instance, if one has the same *totem* as your mother, that person can either become your mother or mother's brother depending on their gender. The resettled people and villagers also trade in various commodities which include crops, vegetables and even exchange seed. In terms of court cases, if the village courts in the resettled villages fail to reach a peaceful resolution one can appeal to the Chief's court, which is located in the communal area. If people feel that they have been tried unjustly they can also appeal to the Chief. However, as mentioned before in chapter 1 sometimes tensions exist between the resettled people and the traditional leadership, particularly the chiefs in the communal areas.

Marriage and kinship[29]

The Shona society is highly patrilineal and its marriage patterns are based on virilocality. Kinship and affinal relations involve social rights and obligations that people have towards specific others in the community and beyond. The following discussion focuses on the two villages in which qualitative fieldwork was undertaken.

A prominent feature of social relations in Mupfurudzi is the intricate pattern of affinal ties indicating intermarriages between people from different kin or lineage groups. In addition, sibling and other cognatic kin bonds play a central role in everyday social life. A similar pattern pertains for the two villages. Other associations are based on joking relationships that do not follow descent lines or affinity, but are a result of friendship pacts between people, usually household heads.

Eleven marriages connect households in Mudzinge (including two in the sample) and seventeen marriages in Muringamombe (including six in the sample). However, during discussions, sample households did not mention these marriage relationships at all except for one household head who was bitter against his son-in-law for causing the death of his two daughters through AIDS. When asked to state sources of information, none mentioned their in-laws as sources of information on maize and agriculture, except one who mentioned an in-law living in another village.

The reason for this reticence to emphasise such marriage bonds seems related to the social rights and obligations that people have towards their in-laws. To minimise conflict between in-laws, and as a sign of respect, in-laws are expected to maintain social distance. Contact between in-laws is limited and they must maintain an air of aloofness. Although intermarriages are functional to the extent that they may further social cohesion, they can also restrict the flow of information within the community.

---

[29] Some of the information on kinship is drawn from notes by Marlene Dekker, an economic anthropologist who also worked in these resettlement areas. See also her thesis (Dekker 2004).

When some people were asked why they had to go out to get information on agriculture instead of asking for advice in the village, they pointed out that people in the village were jealous or suspicious and were often stingy with their information. Even when it came to working in other people's fields in return for money or maize, they preferred to look for work outside the village. It is highly plausible that these elements can be explained by the sometimes tense relations between in-laws.

In Mudzinge, three household heads had their siblings living in other households in the village. When questioned, one of the respondents did not even allude to this relationship, as there was little contact and communication between the families except in cases of emergency. The other respondent had two brothers living in the village but he maintained little contact with them as he himself lived at his field. The third sibling relationship was characterised by open conflict as both spread malicious rumours about each other to the extent of reporting each other to the resettlement officer. In Muringamombe no such relationship existed in the sample households.

Moreover, although some people mentioned relatives as a source of information about the new hybrid seeds, they hardly ever mentioned siblings as a source of information. However, two people in the sample admitted to obtaining agricultural information from their brothers who lived in other villages.

Relationships that seemed to be most enduring and that enabled the exchange of information were those based on *usahwira* – formal joking friendships. These are institutionalised friendships, formally relating to funeral services between families, and involving a frequent exchange of services and gifts (Bourdillon 1987: 61 f). Out of the seven people in the sample in Muringamombe, six were involved in such relationships. These relationships are relaxed in nature, and it was mostly through these relationships that information was spread, advice passed and resources distributed within these communities. Such joking relationships are taken up by choice (rather than through kinship, which is ascribed) and provide a system of support and security in times of need.

Karidza, Chinakidzwa and Chapinduka had the same totem, but their relationships were based more on joking friendships. Karidza maintained that he depended on help and advice from Chinakidzwa and Chapinduka. When asked to mention sources of information on new hybrids people tended to mention joking relationships both outside and within the village more than any marriage or affinity bonds.

Kinship bonds were fraught with difficulties because of jealousy and suspicions of jealousy. Among the Shona it is also believed that only your relatives can kill you through witchcraft means. Strangers are not able to do that. Thus, one can never trust one's own kin. This could explain why kinship bonds are not strong.

Institutions[30]

AREX (Agricultural Research and Extension)

AREX is a department that emerged from the merging of the Department of Research and Specialist Services (DR&SS) with the extension and cropping part of AGRITEX,[31] following the dismantling of AGRITEX in 2002. AREX was briefly known as DAREX; until in 2001 it was re-christened AREX.[32] Although it underwent many changes its main focus remained on research and extension. Asked whether the change from AGRITEX to Darex had affected their operations, an AREX officer pointed out that,

> Danda (the then deputy director technical of AGRITEX in charge of the reorganisation of AGRITEX in collaboration with a German consultant by the name of Conoley) caused those changes. At that time, we began to work as teams. Each group was composed of five to six people and the group leadership was rotational. He encouraged specialisation: for example, we had a grain specialist, tobacco specialist, livestock specialist (pen fattening) and horticultural specialist. One person would operate over a very large area focusing on his or her area of specialisation. The way we operated became very different. I was now operating from the resettlement areas to Bushu. My area of operation ended at Chevakadzi School. When Danda died, we went back to the old system. Maybe he had observed the system somewhere in Europe where he usually visited but that system was not working here.
>
> *You said you used to work here with other officers but now you are all alone: what happened to the others?*
> AREX is suffering from a critical shortage of staff. Most of the officers went to the Fast Track resettlement because there is a need for trained people to educate the new farmers. I also wanted to go to the Fast Track resettlement but then I realised that I would face accommodation problems.

The government still depends on trained extension workers as a way of disseminating agricultural information to farmers.

Bolding (2004: 82) points out that the overall objective of AGRITEX was 'to implement the agricultural policy of government through the provision of agricultural technical and extension services, which stimulates the adoption of proven agricultural practices leading to increased, sustained and profitable production' (see also Mutangadura 1997: 34). True to its stated mission,

---

30 Note that the categories of people patronising any one institution are not exclusive.

31 For an explanation of the formation of AGRITEX in 1981 see Zawe (2000) and Bolding (2004).

32 Because of the confusing change of names sometimes I will refer to AREX as AGRITEX because during the first year of my research the AREX was still known as the AGRITEX. Thus the use of the name will just indicate the time when the data was collected.

AREX (AGRITEX) has been an important mechanism through which technical information regarding agriculture is passed onto farmers. AGRITEX staff played an important role in the dissemination of hybrid maize in the immediate post-Independence period. The role of AGRITEX was to teach farmers to adopt the results of research from agricultural research institutions into their farming practices. Mutangadura (1997: 35) says that in 1989 AGRITEX had a staff component of 2500, of whom 1600 were extension workers, whereas Murwira *et al.* (2001: 302) estimate the staff component of AGRITEX in 2001 to be 2000, although they do not make it clear how many of these 2000 people were doing the actual extension work.

According to the AREX officers, the role of extension officers has not shifted much with the renaming of the department as AREX.[33] Mr Nyamaharo, an officer in the department of AREX, briefly explained the role of the organisation as follows:

> Mostly we play the advisory role. We assist farmers with their operations. For example, we help them in choosing crops, general crop management. We also provide services like soil conservation, water conservation pegging of dams. In general I can say we are into extension.

However, Mr Zawe the Chief Irrigation engineer in the department of irrigation (formerly the irrigation division of AGRITEX but now operating as a department on its own) states that this is a misunderstanding on the part of AREX officers. The change from AGRITEX to AREX was more than just a name change. According to him, the operations of AREX are now limited to research and extension but the practical side of agriculture has to be left to the various practitioners.

For example, the pegging of contours, which was performed by AGRITEX should now be done by the Department of Agricultural engineering (formerly the division of soil and water under AGRITEX). The AREX officer can only inform the relevant department of the need for contours in a certain field but does not peg the contours anymore. For him the confusion is caused by the fact that

> After the dismantling of AGRITEX the ground staff did not change from the District level downwards. They did not have workshops to tell the extension workers of the changes. That is the source of the problem. At the moment there is no department that is well established at the district level like AREX. AREX is more prominent on the ground. Some people in the AREX department still think that things are going to go back to the way they were before AGRITEX was

---

33 For a description of the new roles and functions of AREX and the several departments emanating from the dismantling of the AGRITEX, see Zawe (forthcoming).

dismantled but that is not going to happen. The changes are advanced but AREX officers can still not understand that someone can just take on their roles, that is, the roles they used to perform when AGRITEX still existed.

AREX is central to the dissemination of knowledge and information in the area. However, as we will see later, AREX is withdrawing from the farmers and the farmers presently do not see AREX playing an important role in the dissemination of knowledge and information. There is also a strong realisation among AREX officers that they do not have anything new to offer farmers.

For the government, agricultural extension officers are an essential component of rural agriculture since they are the ones who are in the field to advise the farmers. AREX is different from all the other institutions in the area because its function is not to make profits but rather to advise and assist where possible. The officers are also always in contact with most of the other institutions in the area because, in order to advise appropriately, they need to know what seed varieties are on the market.

The AREX also holds classes where people are instructed on farming issues. These classes are usually held at Ponesai Vanhu Technical College (a couple of kilometres from Madziva Mines) where other interested parties such as the Forestry Commission and CAMPFIRE also come to teach farmers about soil, tree and animal conservation.

AGRITEX and farmers

Agricultural extension officers employed by AGRITEX perform a number of roles including running periodic courses (including those leading to a Master Farmer's Certificate), holding field days prior to the planting period, visiting farmers' fields, and, as described above, liaising with institutions such as Seed Co.

AGRITEX officers impart general knowledge on agriculture through periodic courses of formal lessons for Master Farmer's certificates and through field days. Women, whether household heads or wives of male heads, usually do not attend the courses because of their high illiteracy rate: none in Mupfurudzi had ever attended the lessons offered by AGRITEX officers. Female household heads do, however, occasionally send their adult sons to attend these courses and pass on the knowledge to their mothers. Also women may informally acquire the information from friends whose husbands do attend.

The relationship between AGRITEX and farmers in these resettlement areas is complex and, at times, contested. Many farmers complain that they can only talk to the AGRITEX officers at village mass meetings. The major grievance against AGRITEX officers is that they do not offer personal attention to individual farmers and that they are thought to offer attention only to the rich farmers, particularly to farmers who grow cash crops. While the poor were convinced that AGRITEX officers only visited the rich farmers, the better-off farmers were also worried that AGRITEX officers were not doing their job properly. Of the better-off farmers in the Mupfurudzi qualitative sample, only one had contacted the AGRITEX officer on an individual basis in the previous year, and this was because he had taken the trouble to invite the officer to his field. Furthermore, at village agricultural meetings, or field days, officers discourage farmers from saving seed and using Open-Pollinated Varieties, although they do advise farmers with regard to saved seed when they cannot afford to buy genuine hybrid seed.

Mrs. Tapfumaneyi (the then acting chief of crops for AGRITEX[34]) said that there is no policy governing the activities of field extension officers. There is no stipulated number of times that extension officers are supposed to visit farmers in a day, week, month or year. She said that officers are usually supplied with motorbikes but with very little money for fuel. Some officers use their own fuel with a view to claiming their expenses back but this is such a cumbersome process that many do not bother to do so. Because of a lack of policy, there is no way to judge the performance or non-performance of extension workers. This might lead to some lazy officers not attending to their duties properly. Mrs. Tapfumaneyi said that it was left to individual officers to come up with their own strategies suited to their area of operation. This is somewhat different from the colonial approach to extension staff. Bolding (2004) documents that there was close supervision of extension staff during the colonial era such that any extension worker who was not seen to be doing his work properly or who failed to hold a successful field day would be fired or moved to another area, depending on the circumstances.

On the contentious issue of whether AGRITEX was carrying out a deliberate policy to concentrate on the good or rich farmers, or on those growing cash crops, Mrs. Tapfumaneyi said recently, this had indeed been the case. She said that in the 1980s the objective of AGRITEX was to bring everyone on board. But it emerged that there were some farmers who had opposed to adopting practices that came with the whites: instead they preferred the traditional way of doing things. These farmers resisted recommendations and stuck to the Open-Pollinated Varieties, cattle manure and little crop-spacing. The opposition of some farmers was strengthened by isolated cases of extension officers having relationships with married women. On the other hand, other farmers embraced extension recommendations and did very well by way of higher yields, increased livestock and better farming practices. Naturally, a good relationship was established between these farmers and AGRITEX. Occasionally these were the kind of

---

[34] The interview was carried out by Pedzisayi Mangezvo in 2001 for the International Food Policy Research Institute project before AGRITEX was dismantled.

farmers who attended farmer-training programmes, field days and field observation trials run by AGRITEX. Because of their knowledge about what was going on in the market, these farmers did well and ended up with more income than others.

The current strategy is that farmers growing cash crops are AGRITEX's first priority. This is reflected in the manner in which the field officers are being deployed. Specialists in paprika, for example, are deployed in areas where the crop is grown or can be grown. The same goes for specialists in tobacco, cotton or grain. The new strategy of prioritising cash-croppers comes in the wake of the realisation by government that smallholder commercial crop production has been neglected over the years. So, indeed, cash-croppers are now considered AGRITEX's first priority. But even impoverished farmers who are willing to improve themselves are very much part of AGRITEX's plans. Here, being 'willing to improve themselves', means being willing to adopt the 'expert' recommended technology packs. As part of its pro-poor approach AREX has also decided to pay more attention to Open-Pollinated Varieties, which many poor farmers tend to favour because of their low production costs; therefore it is part of a government committee addressing this issue.

Although poor farmers applaud the focus on Open-Pollinated Varieties of maize, the shift in AGRITEX priorities, together with the fact that AGRITEX predicted drought in the 1999/2000 season, when in fact there were heavy rains in January and February, combined with the current political climate, has led to an increasingly mistrustful atmosphere. Some Mupfurudzi farmers view the phasing out of the older 'more reliable' varieties and replacing them with the newer 'less reliable' varieties as a conspiracy between AGRITEX officers, whites and Seed Co to discredit the government.[35] While not everyone trusts AGRITEX officers regarding seed choices, their expertise is often acknowledged in the case of crop diseases and their control. It is easy to treat a disease when you apply the right chemicals and the results are immediately visible and attributable to the medicine. On the other hand, the climate is unpredictable and could spell disaster for people if the choice of variety was based on an erroneous prediction.

Other factors affecting the ability of AREX to disseminate information include re-organisation, demands placed upon it as a result of Fast Track land reforms, and HIV/AIDS that has resulted in a very high rate of mortality amongst field extension officers.

---

[35] This should be viewed in the present context of political uncertainty.

Seed Co-operative Zimbabwe (Seed Co) and AREX[36]

Seed Co is the dominant institution involved in the production and marketing of maize seed in Zimbabwe. Its seeds were widely used in the study villages. Seed Co works with AGRITEX/AREX. Seed Co is Zimbabwe's largest seed producer. It was formed in 1932 and has been involved in seed production since 1940 (Seed Co 2004). Seed Co started with manufacturing an Open-Pollinated Variety in the 1940s and in the 1960s it produced its first hybrid SR52. Seed Co is mainly involved in research such as seed-breeding and it sees its role as producing quality seeds for farmers. Although in its initial stage it focused mostly on maize which was becoming the fastest-growing food and cash crop at the time, Seed Co later moved on to researching other crops such as Soya beans, Wheat, Barley, Groundnuts and Sorghum.

Mrs. Tapfumaneyi, the Acting Chief of Crops for AGRITEX, was very clear on the links between her agency, Seed Co and other seed companies. Information relating to new seed varieties is passed on from seed companies to AGRITEX, whose field officers in turn disseminate the information to the farmers, by word of mouth through field days. The Seed Co-operative informs AGRITEX that a new variety has been bred. After learning of the basic features or characteristics to expect from a crop grown from that seed, AGRITEX carries out 'field observation trials'. The trials are not carried out at a plot owned by AGRITEX; nor are they conducted prior to marketing the new varieties to the farmers. Field officers run the trials on a piece of land owned by one of the farmers in the community. These on-farm assessments are meant to benefit not only AGRITEX, but also farmers. These trials are used to make area-specific recommendations. For example, when the maize breeders recommend a particular fertiliser, AGRITEX may end up recommending manure from cow dung or dried tree leaves (*mupfudze*). A late-maturing variety in a drought-prone area may lead to AGRITEX recommending early planting. A variety that is prone to grain borers such as R201 may lead to AGRITEX recommending the use or non-use of crop rotation. Efforts are made to adapt the new seed technologies to local use, otherwise, according to Mrs Tapfumaneyi, 'no farmer is going to make use of such technology'. AGRITEX usually relays some feedback to the seed companies but no collaborative research activities are underway or envisaged. The relationship between farmers, AGRITEX and the Seed Co-operative of Zimbabwe indicates that farmers are not consulted on anything but are expected to adopt the products of research wholesale. Even when AGRITEX recommends to suit local conditions, farmers are rarely consulted as AGRITEX recommendations are based on their own scientific knowledge of the agro-ecological zone in which they are stationed.

---

[36] Apart from seed companies, AGRITEX also interacts with chemical and fertiliser companies. The nature of interaction varies. With fertiliser companies, it is usually to get a price list for fertilisers.

While seed companies look at the general characteristics of varieties – taste, drought tolerance, storage qualities, trade value, drying rate after harvest, etc., it is the role of AGRITEX to adapt the new seed technologies to local use or to point out to farmers the varieties that are most suitable in different instances. Mrs. Tapfumaneyi mentioned the examples of SC501 and SC513. SC501 is susceptible to grey leaf and the variety is being worked on to address that weakness, which accounts for the development of SC513. Field extension officers are therefore expected to explain to farmers that SC513 is just an improvement of SC501. If one is in an area where grey leaf is not a problem, then one can grow SC501. If one lives in the same area but has a plot with deficiencies that culminate in the grey leaf, then SC513 is the right variety to plant.

Seed availability is not determined by AGRITEX and seed companies do their own marketing through various promotions. Mrs. Tapfumaneyi said that agro-climatic factors account for certain maize varieties being available or not in certain areas. AGRITEX, however, has no particular policy on the availability of seed in different areas. Mrs. Tapfumaneyi commented that AGRITEX has no policy governing Open-Pollinated Varieties (OPV), old varieties and saved seed. She reiterated that the discretion lay with the farmer to pick on a variety best suited to his/her plot. In Mupfurudzi, some farmers claimed that AGRITEX was conniving with seed companies to market 'bad' seed, an allegation denied by Mrs. Tapfumaneyi.

GMB (Grain Marketing Board)

In this section, I do not consider the various criticism on the failures of the grain marketing board, nor do I discuss its successes. I consider the GMB in relation to how the farmers in my study area interacted with the organisation. Jones (1987: 375) correctly points out that in tropical Africa, agricultural marketing boards are 'heirlooms of the great depression and World War II, when colonial governments found their principal sources of revenue severely reduced and both European and African populations financially distressed'. In this vein, the Grain Marketing Board in southern Rhodesia was established in 1931 in response to the 1930 world recession. It has a mandate from the days of its inception to ensure food security in Zimbabwe with particular reference to staple food products, namely, maize and wheat. It also has to guarantee the orderly marketing of agricultural products, mainly grains, oilseeds, edible beans and coffee, within Zimbabwe. The GMB buys grains from farmers and sells them to the domestic agro-processing industry in addition to exporting these products to regional and international markets. In cases of food shortage, the GMB has the mandate to import from other regional and international markets (www.dexelzim.co.zw).

In the research area the GMB is located at Tafuna, on the road to Bindura from Shamva, although recently it has opened a depot in Muringamombe. GMB offers marketing services as well as loans to farmers. It offers a higher price than the other buyers in the market. This price incentive, however, is eroded because people have to pay transport costs and they have to pay for the use of the bags, and payments are delayed. For example, in the 2000-2001 season, farmers were paid their money after the 2001-2002 season had started. In the *Murimi Wanhasi* ('Today's Farmers' / 'The New Farmer') Programme on Television (22 January 2004), the farmer of the week complained that, although the new farmers had sold all their maize to the GMB in the previous season, they had not yet received their payments.

GMB was very popular among farmers because of its seed and fertiliser loans scheme for farmers. Although the GMB did not hold any classes for the farmers, it was very central to the dissemination of knowledge and technology among the farmers. For instance, some farmers justified the amount of fertilisers they used per acre by using the amount of fertilisers they got from the GMB as a standard measure. Mr. Chari, a farmer in Muringamombe, said that he had used 5 bags of fertiliser per acre, 2 bags of Amonium Nitrate and 3 bags of compound D because that was the amont of fertiliser he got on loan per acre from the GMB.

The GMB was also important for the dissemination and adoption of new hybrid varieties of maize. Especially farmers who depended on the GMB for their seed and fertiliser needs, would plant any maize variety that the GMB provided them with. Thus it is hardly surprising that farmers were often not happy with the maize variety given to them but they were powerless to do anything about it. Baba Peter, a poor farmer in Muringamombe, was one such person. He had planted a late maturing variety (Sc735) late in the season because that was the variety he got on loan from the GMB. The GMB was late in disbursing inputs to people. According to him, everyone who planted the SC735 was disappointed because the seed did not do well.

In a similar vein, although farmers still wanted the old maize varieties, they had to adopt the new maize varieties because those were the ones they were getting on loan from the GMB. Another farmer commented that:

> 401 is useless. No matter how much fertiliser you apply, the maize cob is small. When selling, it is very difficult to get a grade A for 401. We want R215 but we can't find it in the shops ... R215 is a very good seed. Even if you plant late, you will get something unlike these new varieties. However, we are getting 401 from the GMB. What else can we do?

Thus, sometimes new knowledge and technology is imposed on the farmers and they simply have to do the best they can.

The marketing strategies of the GMB have also shaped the adoption of new and the abandoning of old technology and ways of doing things:

> We also used to grow *mapfunde* (millet) but we stopped because we had marketing problems. In 1987, we got 36 bags of *mapfunde* and it was fetching

> $500/tonne and maize was $250/tonne. However, only a few of us grew *mapfunde* so it became expensive for individuals to transport it to the GMB. The other obstacle was that the market for *mapfunde* opened up after the maize had been sold. By the time the market opened, most of the millet would have been lost to weevils. As a result, we shifted from *mapfunde* to maize.

During field work no household in the sample cultivated millet, although most households maintained that they cultivated millet before they were resettled.

The Cotton Company of Zimbabwe (Cottco)

Cottco's mission statement is to 'facilitate the most profitable and efficient growing, processing and marketing of agro-industrial products to the benefit of all stakeholders' (Cottco 2004). As well as marketing, Cottco provides farmers with agronomic and financial support at every stage of the production process. Cottco is an off-shoot of the Cotton Marketing Board, which was formed in 1969 to oversee the handling, marketing and processing of cotton. In 1994, Cottco was incorporated and then finally privatized in 1997. Now it is run as a commercial company and it has also acquired 34.75 per cent of shares in Seed Co. Cottco is not only a major player nationally, but is also a major institutional player in the area of the research. In Mupfurudzi, Cottco buys cotton and offers seed and fertiliser loans. Its current competitors are Cargill[37] and Farmers' World. It employs a grading system for giving people loans. People in the gold group (those reaping more than 3,000 kgs of cotton in a season) obtain as much as they want, while those who reap less have to resort to group loans. Farmers also use fertiliser from cotton loans in their maize fields. In the sample seven people were involved with Cottco.

Although Cottco is primarily concerned with marketing, it is an important knowledge-disseminator for cotton farming. Cottco sponsors field days for farmers where farmers learn about cotton farming. Cottco selects capable farmers and provides them with some inputs. If the crop has done well, they hold a field day, to which all farmers are invited, to observe the crop and hear what the farmer has done to produce such a good crop. When asked about the role of Cottco in knowledge dissemination, Mr. Mushayi Mapeto, the Cottco Collection Point Supervisor (CCPS), pointed out that the Cottco advises farmers on how to take care of their crops, especially how to guard their crops against pests and diseases and how to offer appropriate treatment if the cotton is attacked by pests and diseases.

---

[37] In 1996 Cottco sold its Tafuna ginnery in Shamva to Cargill and hence farmers who want to sell their cotton to Cottco have to go to Bindura, about 60 kilometres away.

The cotton company has linkages with AREX and so AREX officers were invited to all the field days I attended and were called upon to participate actively in the proceedings.

Apart from organising field days, Cottco officials are more proactive in their dealings with farmers. Cottco gives fertiliser and seed loans to farmers with the aim of making profits. They can only obtain these profits if farmers are able to farm productively. Cottco has to make sure that the fertiliser it supplies is used for cotton farming and not for other crops. As a result they have to carry out field visits to examine crops and advise the farmers wherever possible.

> Cotton speaks for itself. If a farmer uses the fertiliser for other things I will know when I go for field assessments because the cotton plant will show that it lacks fertiliser. On top of that farmers who do so cheat themselves because cotton is a very jealous crop. If one diverts the fertiliser, the yield will not be very good. To make a profit from one hectare one should get between 12 to 15 bales and above. Below twelve there is no profit. If you do not apply 5x 50kgs of compound L per hectare the cotton flowers and balls will fall off. If one gets 20 or 15 balls per plant then that means a terrible loss. One should get between 60-150 balls per plant.

Although field assessments were not very popular among farmers who diverted fertiliser that was meant for cotton, to maize, the field assessments gave them the opportunity to learn more about cotton on a one-to-one basis with the Cotton experts.

Cargill Zimbabwe

Cargill is an international seed company operating in 59 countries. In Zimbabwe it owns three cotton ginneries, one of which is located in Shamva at Tafuna. Cargill processes 30 per cent of Zimbabwean Cotton (Cargill 2004).

Cargill is the Cotton Company's largest competitor. However, apart from buying cotton, Cargill also buys maize from farmers. When it comes to marketing Cargill is preferred because it gives people cheques 4 hours after the initial sale, while in the case of Cottco there is a 24-hour delay in payment. Its only disadvantage is that it does not offer any loans or back pays (bonus cheques) like Cottco does. In the sample two people had dealings with Cargill, though I have reason to think that maybe this number is not correct since a lot of people came to obtain Cargill cotton bales from Mr. Karidza, the Cargill representative, when I was interviewing him.

Although Cargill does not offer any loans to farmers it also plays a major role in disseminating information to farmers. When asked about demonstration plots the AREX officer Mr Nyamaharo mentioned Cargill as one of the companies that sponsored field days.

FSI Agricom

FSI Agricom is an indigenously owned company, which became fully operational at the end of 2001, and couched its objectives in highly political terms. Its CEO was quoted in Newsweek (2002), saying: ‘The perception outside is that whites are the ones who make Zimbabwe tick ... we are creating a centre of excellence to demonstrate what is possible’. Thus FSI Agricom intended to give the new farmers (under the Fast Track Land Resettlement programme) seed, ploughs and other inputs and equipment on loan and to subsequently market the resultant crops. However, FSI Agricom’s scope of operation has not been limited to the ‘new farmers’ but has extended to others in communal and early resettlement areas. The FSI Agricom also had its own farms, where it intended to carry out its own farming activities and venture into agro-processing. However, in 2004 four of its farms, totalling 4 300 hectares, were listed for acquisition by the government under the Land Acquisition Act. When this happened its focus shifted slightly to cotton procurement and the sponsoring of out-growers (*Sunday Mirror*, 25 July 2004). Agricom opened its field offices in Shamva in 2002. Agricom is a national organisation with offices in Shamva, Bindura, Darwin and Kamutsenzere in Dande, and several other places, mostly in Mashonaland Central. Their headquarters are in Harare. At the time of this research, they were still trying to establish themselves in the resettlement areas and although they had been to Mudzinge they had not visited Muringamombe village.

They were still in the process of selecting village representatives and in Mudzinge had selected Mr Chitabura to represent them at village level. The purpose of these representatives was to work with the farmers and to vet them for loans at the village level so as to avoid accumulating a lot of bad debts. Thus in Mudzinge the Agricom representatives were popularly referred to as ‘Chitabura’s people’. Agricom offered loans to farmers specifically for cotton and maize and, in return, farmers had to sell their crops to them to clear the loans.

Before they could get fully established in the area, the Agricom representative had to obtain written permission from the DA to allow them to start operations. In 2002, to operate successfully in the area, any company had to have the backing of the political leadership and of the top leadership in government. Agricom promised the District Administrator (DA) that they wanted to buy 22 tractors to provide tillage for farmers (for hire) and they wanted the DA to obtain the tractors for them. However, according to Mr. Chinyani from Agricom, these tractors would only till the land for those farmers who had bought more than 10kgs of seed from them.

Although Agricom was similar to Cottco in many respects, there were also some differences. Agricom also supplied ploughs, building materials, and loans to buy scotchcarts and wheelbarrows, and deducted money for payment from cotton sales. If people failed to service the loans Agricom could impound their cattle. Compared to Cottco, Agricom was regarded as better, as they paid a good price for the cotton and did not charge for delivering the inputs to the farmers or for collecting the cotton from the farmers to bring it to their warehouses. Agricom also seemed relatively wealthier as it had better cars than Cottco had.

Agricom was therefore a promising player in the dissemination of technology in terms of farming equipment such as wheelbarrows, ploughs and scotch-carts. Like other organisations, it was central to the dissemination of technology.

Zimbabwe Tobacco Association (ZTA)

ZTA focuses on tobacco production. Most of the information from ZTA concerns forestry, the prevention of soil erosion, and the growing of gum trees. Tobacco curing in Zimbabwe still depends on trees instead of coal. As a result most of ZTA advice focuses on the growing of gum trees for tobacco curing in order to prevent the cutting down of indigenous forests. Mr. Mavheneke Chikerema, a tobacco farmer in Muringamombe who had attended ZTA lessons, had a gum-tree woodlot that he was managing. Asked about the operations of ZTA, Mr. Mavheneke Chikerema replied:

> I cannot say they just focus on tobacco because they look at a lot of things. For instance, they are also concerned with the prevention of *veld* fires. They do not like the destruction of grass or of trees in the environment. They say that they focus on tobacco but they really deal with the farmer in general. For instance, they look at issues concerning the farmer such as how to deal with pest and diseases. How to prevent aphids and how in general to control pests. All that knowledge is given by ZTA. (*he thinks for a while, then says ...*) 'Crop rotation' they teach us about that. They say tobacco can not be planted twice consecutively in the same field so that we have to rotate the crop. One year you have tobacco, the next year you plant maize in that field and the third year you can leave the field fallow rather than cultivate it.

ZTA also works in conjunction with AREX. All 'good' tobacco farmers were said to be good because of their association with ZTA and AREX.

Agriculture Finance Corporation (AFC)

Although it is now officially known as Agribank, farmers still refer to this organisation by its old abbreviated name, AFC. During the colonial era AFC was known as the Land Bank. The Land Bank had its headquarters in Bulawayo and was operated by the British South Africa Company to provide credit to farmers. When self-rule was introduced to Southern Rhodesia, a bill was introduced authorising the government to raise one million dollars to take over this bank and establish its own with a capital of six million dollars.

In 1971, the Land Bank was officially changed into the Agriculture Finance Corporation (see Smith 1985 for a history of the AFC.)

The Mugabe government, after introducing the Fast Track land resettlement programme, wanted to extend lines of credit to the new farmers, who more often than not did not have collateral. As a result, the AFC was restructured to become the Agribank which could extend loans to farmers without the strict requirements of the provision of collateral. Within Agribank and government circles it was clearly realised that it was important to extend loan facilities to aspiring farmers, to enable these farmers to implement their farming activities successfully (Chakwera 2002). In the 2003-2004 season, the government committed $Z15,000,000,000 to Agribank for this purpose. In the past people got loans to buy fertilisers, seeds and cattle from the AFC. However, currently Agribank focuses only on tobacco farmers, though the tobacco farmers may also use this money to buy maize inputs. In the sample, only one person, a tobacco farmer, had any dealings with the AFC.

Farmers' World

Farmers' World is a privately-owned organisation that came into existence as a result of trade liberalisation. It has been in operation in the area for two years. It mostly specialises in the buying of cotton and maize from farmers at competitive rates, and also provides seed and fertiliser loans. Individual farmers are supposed to pay back their loans by selling their crops to Farmers' World. From the crop delivered by each farmer, Farmers' World deducts the amount of crop equivalent to recover the loan. The individual farmer is paid for the remainder of the crops after the necessary deduction. This is viewed by many farmers as a kind of 'contract farming'. However, individuals need to pay a fee of $Z250 to be considered for the loans scheme. Four people in the sample had done business with Farmers' World. At the start of the selling season, it usually pays more than the GMB rates.

Purity

Purity is a milling company located at Madziva Mine. It is solely a trading company owned by an indigenous business person. Purity specialises in buying maize from farmers at competitive rates. Although it pays $1000 (quoting 2001 prices) (US$18 official market and US$1.60 illegal parallel market) lower than the GMB price for grade A maize, Purity is an advantage for the less confident farmers in that the maize is not graded and all maize therefore fetches the same price. Purity has a further advantage in that people do not have to foot transport costs since they can deliver the maize themselves using scotch carts. About six people admitted to having sold their maize to Purity.

Purity provides no other services besides the marketing and milling of maize. It does not offer any loans. Some farmers sold their maize to Purity as a way of evading loan repayment to GMB, Cargill, Agricom or Farmers' World. Furthermore, usually it is the women and very poor farmers who sell to Purity. The rich farmers generally sell to Purity only when they urgently need money.

Religion

There are two dominant religions in the area, based on Shona and Christian beliefs. I am loath to call Shona religion 'traditional' since this does not capture its essence. As shown in Chapter 5, various interpretations of *Chisi* point to the fact that Shona religion is indeed more fluid, and less determined and hierarchical, than the word traditional might suggest. The second religion is that of the Christian faith. Although these religions differ in their beliefs and practices it is of course possible for people to follow the teachings of both or select various attributes of each religion that make sense to them.

Shona religion

There is no one coherent religion among the Shona people. The Shona consist of five major groupings: the Karanga, Ndau, Manyika, Zezuru, and Korekore peoples. Despite the socio-religious variations among the Shona there are also commonalities. Since I worked in a resettlement area made up of people from different regions of Zimbabwe, it is impossible to adopt the concept of a 'local' or 'indigenous' religion. In this section I therefore identify the similarities and major variations in the religion of the Shona people. The resettlement area I worked in fell under the Korekore people of the Nyamaropa Clan, whose totem is the eland (*Shava nematombo*). Hence any display of Shona religion in the public sphere at the community level was guided by beliefs associated with the Korekore of the Nyamaropa *shava nematombo* clan.

Another reservation against using the term 'Shona religion' is that it is questionable whether Christianity is not also Shona religion. Johanne Marange and Johanne Masowe started their Christian churches in 1932, and over the years Christianity has come to play a crucial role in Shona everyday practices and in this sense is thoroughly Shona. Nevertheless I wish to heuristically separate Shona religion and Christianity. Firstly I differentiate between these two religions on the basis that Shona religion was the norm among the Shona before the coming of the white man. Thus, although Christian values have been internalized by many Shona people, from this angle Christian religion is relatively foreign to Shona people. Secondly, although Christianity has also been Africanised differences continue to exist between Shona religious beliefs and those espoused by Africanised versions of Christianity. The third reason for making this distinction is based on the fact that the people themselves distinguish between *chivanhu chedu* (our religion) and *chikristu* (Christianity).

Shona religion can be understood at both the family and community level and is characterized by honouring the dead either as *vadzimu* (family ancestral spirits) or *Mhondoro* (lion spirits/clan spirits); at the apex are *Mwari*[38] (God), *Musikavanhu* (the Creator of People), and *Nyadenga* (the Owner of the Skies), depending on which group of the Shona people one is referring to. The Karanga and the Zezuru and the Manyika do not talk of *Mwari* but of *Nyadenga* (the Owner of the Skies) and the Korekore speak of *Dedza* (the Rain Giver) or *Musikavanhu* (the Creator of People) whilst for the Manyika *Musikavanhu* is the name of a powerful ancestral spirit. Other names that are used by the Shona to refer to God include *Matangakugara* (the One who was There First), *Samasimba* (the All Powerful One) and *Zame*. The common trait among the Shona is that people do not pray to the ancestors but ask them to mediate between themselves and the Supreme Being who is referred to by the various names just alluded to. The names simply refer to the various attributes of God and the name used depends on the subject under discussion. For example, when talking to God as the giver of rain people refer to God as *Dedza* (the Rain Giver). When talking of the power of God people refer to God as *Samasimba*. Difference arises only in the style of worship since each group has its own ancestral spirits that mediate between themselves and God. These ancestral spirits can demand to be honoured in a variety of different ways. At the base there are God's creations, including people and animals; between God and the living there is the country of spirits (*Nyikadzimu*) where all the spirits of our dead ancestors and relatives reside. Mararike points out that 'Shona religion is divided into three departments. The first department is that of God (Mwari). The second is the department of spirits (*mudzimu*) and the third department is that of the living people'. The problem with this conception is that it gives the impression that there are clearly demarcated boundaries between the living, the dead and God, when in reality boundaries are often blurred. The following quotation from a villager cited in Mararike (1998: 160) highlights why I have reservations:

---

[38] Daneel (1971: 81) defines *Mwari* as a High God who seems perhaps less directly involved in people's individual lives, as compared to the ancestors, but who who can be consulted in matters of national importance. As a result of their individual problem-solving at the family level, people in African religions depend mostly on their ancestors instead of on Mwari. Mwari is responsible for the fertility of crops and, 'as the God of the fertility of crops, Mwari is first and foremost regarded as the rain giver' (*ibid.*); however, Mwari can be approached only indirectly through senior lineage ancestors. Among the Shona in Zimbabwe, whether they are Christian or not and regardless of their ethnic group, everyone now uses the term Mwari to refer to God.

> The dead are dead but alive. They do not see but they see. They do not hear but they hear. They do not walk but they travel long distances. The dead are away but with us. The dead protect us but need us. We need the dead but the dead need us. Without us the dead are really dead and without the dead the living are in the jaws of death. The dead are our most valuable assets. They give us all other assets.

Hence departmentalisation would suggest a clinical neatness and would lead to the compartmentalization of phenomena that are very complex. Relationships between the living and the dead and God are so intimately enmeshed and intertwined that they defy departmentalisation and compartmentalisation. Maybe it is more apt to talk of levels in one big whole, where at the top we have *Mwari*, in the middle there are *midzimu*, and at the bottom, the living and those who are yet to be born.

After their death people aspire to live with their relatives in *Nyikadzimu*. However, although this is desirable it is not possible for everyone. All those who have died unhappy or who had strong grievances[39] against some living persons, or who were murdered or whose death rituals were not properly followed, form a 'community of wanderers'. All those who still have grievances against the living or who did not die a 'proper death' (natural) will have to come back to haunt the living until their grievances are addressed. Only then can they claim their place in *Nyikadzimu*. Those who were murdered will come back to haunt the families of the murderer, causing havoc, death and untold suffering. The avenging spirit is normally referred to as *Ngozi* .[40] This spirit can rest only when the family of the murderer pays reparation to the family of the murdered person. After the reparation is paid the relatives of the murdered person in *Nyikadzimu* can welcome their relative to the other world.

God and *Mhondoro* are relevant at the community level. A common trait among the Shona is that their religion is not composed of constant prayer and supplication to God and the ancestors. People only offer supplication and prayer occasionally, and they are usually accompanied by the brewing of beer. People brew *doro remusha* (the beer of the family), then drink, dance

---

39 An example of a grievance that would cause a person to come back as an avenging spirit could be a mother who was beaten or ill-treated by her children when she was alive. The spirit can cause the deaths or children and grandchildren to be generally unlucky. For example, it can result in a failure of the children and grandchildren to get married or, when they do, to have unsuccessful marriages.

40 However, it should be noted that the term *ngozi* is used differently by different cultural groups in Zimbabwe. Bourdillon (1987: 268) points out that among the Korekore and the Tonga, *ngozi* can refer to friendly family ancestors that possess mediums. In his study of livelihoods in Buhera, Mararike (1998: 160) uses the term *ngozi* to refer to 'angry spirits'.

to traditional music, and perform rituals. This *doro remusha* functions to celebrate and thank the ancestors for their blessings. There is also *doro regoho* (the beer of harvest), when people thank the ancestors following a good harvest. During the *doro regoho* festivities the Shona take token quantities of their crop harvest to the *Mhondoro* or other spirits of the land to show the spirits that they had a bounty harvest and to to thank them. If things go bad the Shona pray to God and/or carry out rituals to appease the ancestors so that whatever was wrong can be righted. Thus, when there is drought among the Korekore they appeal to the *Mhondoro*. However, the *Mhondoro* is not the rain giver as such but rather the intermediary between *Musikavanhu* (God the Creator) or *Dedza* (the Rain Giver) and the people. Likewise at family level, for example, when children are not getting married as they should, or whhen there are a lot of mysterious, unexplained deaths, people ask *midzimu* (family ancestral spirits) to intervene.

As Mararike (1999: 72-3) points out, among the Shona, 'before the introduction of Christianity and other foreign religions, there was no separation between religion and other human activities. The relationship between the living, the dead and God was intertwined ... the natural world, the human world and the spiritual world are closely intertwined'. But my view goes somewhat beyond this assertion, since, as shown in Chapter 5, religious beliefs and practices are also embedded in technology and in agricultural activities

Generally, in Shona religion at the family level, totems are important and should be recognised since they demarcate one group of people from another. Totems involve taboos concerning sex and eating. If these taboos are not observed or if the wishes of the dead are not honoured, the spirits of the dead can easily become angered and exact their vengeance on the living. For example, the village head of Mudzinge maintained that a friend of his from another village had fallen seriously ill and a lot of unexplained deaths and misfortunes were happening at his home. However, it later emerged that all these things were being caused by his late half brother. He had told them that he did not want to be buried at the local cemetery but on his homestead. Since resettlement laws did not permit this, the half brother had overruled his late brother's wishes and had buried him at the local cemetery. However, because of a lot of mysterious happenings which culminated within a month of his death in the death of two sons of the deceased and the illness of his half brother, it was later resolved that the dead brother had to be reburied on his homestead as he had wished. The chief and the District Administrator gave permission for the exhumation and reburial to take place although the Chief pointed out that some traditional rituals would have to be performed at the exhumation so that the spirits of the land would not be offended. The dead man was exhumed twenty-eight days after his burial and he had not yet started to rot. The widow of the exhumed man died on the day her husband was exhumed because she had insisted, against advice from the elders, on seeing the body of her husband again. After the reburial things went back to normal and the bad happenings that were haunting the family stopped.

Thus the most important thing at the family level among the Shona is to live in harmony with one's ancestors as well as other dead members of the family. This can be achieved through honouring their wishes and brewing beer once in a while to appease them. Following death rituals properly is also one of the ways in which harmony can be maintained between the living and the dead.

Christianity

There are many different Christian denominations in the area, but the most dominant are those of the Apostolic Faith, especially the Johanne Masowe weChishanu Apostolic Church and the Johanne Marange Apostolic Church. People in the areas preferred these churches because of the faith healing opportunities they present. Both churches have also managed to borrow from traditional beliefs in spirit possession and so people flock to these churches to be delivered from spirits that might afflict them. They also offer the hope of dealing with everyday problems through prayers or receiving holy water from prophets. As noted by Bourdillon (1987: 292), 'they offer material advantages in the supposed healing powers of many of their ministers'. (For more on the reasons why people are attracted to independent churches see Bourdillon 1987; Murphree 1971; Dillon-Malone 1978 and Daneel 1971; 1977.)

Here I will provide a brief account of one independent church, that of Johanne Masowe (John of the wilderness) Apostolic Church, and mention Johanne Marange Apostolic Church only in passing. These two churches had the largest number of followers and their teachings and doctrines had much more impact on knowledge than those of all the other churches. However, I concentrate on Johanne Masowe Apostolic Church since the majority of my informants claimed to belong to it. There was only one member of Johanne Marange Church in my sample.

On the other hand, for the missionary churches there was only the Roman Catholic Church. Membership of this church was mostly limited to Madziva Mines but when the mine closed the church ceased to function. Jehovah's Witnesses and the Apostolic Faith Mission were also centred on the mine and did not have an impact on the thought processes of villagers. During the 2002 elections, the Jehovah's Witnesses did not find favour with villagers since they refused to buy party cards as well as vote in the elections. Referring to the Jehovah's Witnesses, one respondent showed his disapproval of their doctrines:

> These *chitawara* (watch tower) people – I am not sure whether their religion is about arguing with others only. They do not like to work together with other people but they want land. Jews and Israelites are very religious people but they

> still unite to fight for their land. What kind of people are they? They even say they are not allowed to vote!

Johanne Marange started the Johanne Marange Apostolic Church in 1932 after being called by God. On July 17 1932 Johanne heard a voice,

> You are John the Baptist, an Apostle. Go forth and do my work. Go to every country [and] preach and convert people. Command them not to commit adultery, steal or become angry. Baptise people and observe the Sabbath (Daneel 1987: 56).

Marange rejected ancestral worship and the use of traditional medicines. However, unlike Johanne Masowe, Johanne Marange also rejected the use of modern medicine to cure illnesses among his followers, since he believed that only God could cure illnesses.

Johanne Masowe[41]

A man named Shonhiwa Masedza Tandi Moyo started the Johanne Masowe Church in 1932 in Zimbabwe (southern Rhodesia) during the colonial era. He had received visions and dreams from God, which pointed to his ministry as John the Baptist. After being arrested by the colonial authorities for 'walking' without identification papers and being brought before the Chief Native Commissioner, Shonhiwa said

> I really do believe that I have been sent from heaven to carry out religious work among the natives. I think that I am 'John the Baptist' as a voice told me so. No human being has guided me in my teachings. I am only guided by the voice that I heard when I was staying on the hill for forty days. I have heard the voice in my dreams. The voice came to me through a bush that was burning quite near me. When the voice ceased the fire would go out ... I no longer suffer from pains in the head (Dillon-Malone 1978: 12).

This Shonhiwa claimed to have been attacked by mysterious illnesses and to have risen from the dead once after he had died of a mysterious illness. It was after one of his illnesses and some mysterious occurrences that God revealed himself to Shonhiwa and told him that he wanted him to do his work. God instructed him to take on a new name – that of Johanne – as he was John the Baptist who had come to Africa to preach the word of God to Africans (the church of Johanne Masowe has branches in several countries in Africa). Johanne Masowe had a firm conviction that he was chosen by God to lead the African people to salvation and his followers were also filled with

---

41 For a detailed understanding of Johanne Masowe's origins, see Dillon-Malone 1978.

the same conviction. The followers believed that the spirit of John the Baptist had taken hold of Johanne Masowe.

The Johanne Masowe church believed in the ritual purity of sisters who were completely dedicated to God as his wives. The Church observed its Sabbath day from sundown Friday to sundown Saturday.

Johanne Masowe later died in Zambia in 1973 and there was a succession dispute that resulted in the church splitting into several factions. The different congregations tended to congregate around the different powerful leaders in their localities. The leader of the Johanne Masowe Church in Shamva and Mount Darwin is Wimbow, and some people referred to their church as Johanne Masowe Yamadzibaba Wimbow (Father Wimbow's Johanne Masowe).

The church is also constantly changing to suit changing circumstances. When it was formed it dealt with problems of powerlessness in mission churches where blacks rarely got to positions of high authority. Now the church also tackles issues of AIDS and health as shown in the conversation below:

> In this area when a man and a woman fall in love and they want to marry, they can go to Wimbow. If Wimbow shakes their hand, that is all.
>
> *(Christine) It will mean they are okay and God is giving them their blessing?*
> No. It will be a final farewell. If he shakes your hand you will know that you will die. You have AIDS.
>
> *Does Wimbow himself tell them this?*
> He does not say anything but as long as he shakes your hand you will know that you have AIDS and you will die.

Concluding remarks

The foregoing discussion makes it clear that the different official organisations – private or government-owned ones – usually see themselves as disseminators of knowledge whilst farmers, who are hardly ever consulted, are expected to adopt wholeheartedly the recommendations from these centres of 'knowledge,' seeking clarification only on how best to use the technological packages offered.

In addition this chapter has underlined that it is not only at the local farmer level that development discourse is couched in political terms: as demonstrated by the case of FSI Agricom, even corporations can resort to playing the political knowledge game in the interest of profit making. Thus, to understand certain knowledge discourses, the background and context that give emergence to specific discourses has to be critically understood.

It is important to note that, even after resettlement, households do not attain the same economic level. Resettlement does not, therefore, have a levelling influence on social and economic differentiation. While poverty increases for some households, others manage to strengthen their economic positions by acquiring more items of wealth than they had prior to resettlement. As will be discussed in later chapters, sometimes these levels of relative wealth and poverty influence who experiments with what and when. In this way, then, their overall household position can influence the vantage points from which farmers view, consider and analyse different methods of farming, farming knowledge and behavioural consequences.

The importance of friendships in information dissemination has been highlighted. Also highlighted, but to be discussed in more detail in later chapters, was the relationship between agriculture and religion.

4

Farmers' knowledge and sustainable innovation: Experiments and observation

Photo 4.1 Cow dying of black leg. Photo by author.

Introduction

By discussing farmer experiments and observation this chapter seeks to recast and the understand farmers in another light, not as people who merely stick to tradition, empty vessels that need to be filled, but as proactive agents who seek to understand the world around them.

Photo 4.2 Cattle vaccination exercise by farmers in Muringamombe. Photo by author.

This chapter will demonstrate that farmers are very innovative and that 'development' agents should not view them as people who have to be changed from outside, and, even so, are resistant to change. Farmers should be regarded as agents and actors who devise ways to deal with problematic or restricting situations so as to pursue their own projects.

Experimentation

Crops

Experimentation is an integral part of all sciences. Whether one believes that experiments are rigorous exercises in search of the truth, or that they are political rhetoric to support specific paradigms, no one can question that scientists do carry out experiments. At the same time lay people, in this case farmers, carry out experiments. Although their methods can be said to be crude and rudimentary, since no variables are controlled for in the 'true' scientific spirit, who could argue that their methods are less scientific than 'true' scientific experiments? Lay people do not carry out experiments to confirm specific paradigms but to get to the 'truth', as they see it. This is so because the results of the experiments are related to their day-to-day lives. Scientists are interested in the end product such as a maize variety that can withstand certain conditions such as excessive rains, whereas farmers are more interested in whether the products suit their needs, which might not be of interest to the scientists. Farmers may consider things like taste, size or whether the crop variety responds well to manure: where the scientist has a thirst for knowledge, the farmer has a thirst for results. Thus experimentation does not end at the laboratory gate; to some extent it begins at the gate.

For instance, after carrying out extensive experiments with hybrid seed, Seed Co (2000: 12) recommended that in order to select which hybrid to grow, farmers 'first need to establish their mean yield over the last three or more seasons'. The hybrid recommendations are based upon the average yield potential of a field. For example, where the yield potential of a field is 3 tonnes per hectare, a farmer should consider early maturing varieties such as SC401. However, if yields of above three tonnes per hectare are expected, then medium varieties like SC513 would be higher-yielding than early maturing varieties. On the other hand, after carrying out their field trials, farmers do not only consider the yield potential of a field but also other things, such as the taste of the variety when eaten. On being asked if she was going to continue to cultivate Panner seed as well as the SC501, both of which she had experimented with, one farmer maintained that

> The *sadza* from SC501 tastes good. However, the variety we grew this year (*Panner*) does not taste good at all. If a visitor comes and you cook *sadza*, the visitor might think that you had mixed mealy meal with soil because of the dull/dark colour of the *sadza*. Others also say that if SC501 receives adequate rain you get a good harvest because it can have 2 cobs on every maize stalk. However, because of the poor harvest we got, I don't think we will be growing it again.

Although the Seed Co manuals recommended SC501 for the Shamva area this farmer reached a different conclusion from that reached by the experts. She would not cultivate SC501 again because the yield was low and, at the same time, she would not cultivate Panner either because apart from the low yield the maize did not taste good at all. Another farmer who had also experimented with different kinds of maize seed claimed that

> ... when comparing SC501 and R215, SC501 is more nutritious than R215. If you go to the grinding mill with a bucket of R215 and a bucket of SC501 with the intention of having pearled mealy meal you will have less mealy meal for R215 than you have for a bucket of Sc501. This is so because the grains of R215 are much smaller than the grains of SC501. Compared to R215, 501 has got top-quality feed. 501 is like a hybrid. However, R215 and R201 taste better than SC501. Sadza cooked from R215 or R201 is much tastier than that cooked from SC501. Even when roasting green mealies, SC501 is tasteless but with R215 you know you are eating real maize.

This farmer preferred to cultivate SC501 despite its being tasteless because he claimed that he was cultivating for commercial reasons and that varieties like R215 were for smaller and less capable farmers. In this respect scientists and farmers consider different things when carrying out experiments, so that their results are bound to diverge and to lead to different conclusions and different applications.

For farmers there are two kinds of experiments: the first one I will refer to as individual experiments, and the other as individual experiments in collaboration with AGRITEX and Agri-business. Most farmers in the sample carried out experiments before adopting new crops. However, their experiments were not defined in the linear way of scientists. For farmers no precise research tools were necessary and their research methods were very crude. What is necessary is the availability of land and seed. The experiments are usually directed at testing whether the seed will meet certain specifications of the farmer. This led some farmers to experiment with maize seed of the *Shumba* and *Nzou* varieties despite the fact that these were long-season varieties and only short to medium season varieties are recommended for this area. One farmer had this to say:

> I tried to grow SC701, SC603, SC511 then I discovered that 7 and 6 are long season varieties. Because of the short season that we have here, if you grow 7 and 6 you won't get a good harvest. I also experimented with 4141. 4141 is very similar to R215 and R201. It's a complete waste of time. If you are growing for your own consumption maybe they are all right but not if you are growing any crops for sale. If you are preparing these varieties for sale they do not weigh much. On top of that they have an unhealthy dark colour. If you grow R215, R201, 4141 you are certain to get nothing but poor grade. People who stay in the reserves (rural areas) should grow these varieties, not resettlement scheme farmers.

When I asked the AREX Officer on a separate occasion whether he agreed with the assertion of the above farmer that the R varieties were for communal farmers he disagreed:

> I do not agree with what that farmer told you. As farmers, we differ in our times of planting. Seed varieties are not much different. It depends on how that farmer manages that variety. In this region some people plant 701 and 709 and they are doing very well. Even if you go to their fields you will be surprised by the crop but these varieties are not recommended for this region since they are long season varieties.

What is interesting in this case is that the farmer did not bow down to the argument of the authorities, in this case the AGRITEX officers, and behind them the scientist who had manufactured the seed and tested it under rigorous conditions. Farmers should be regarded as active in the production of knowledge as they do not simply equate knowledge with professional and specialised ideas and information, but augment this with the knowledge they generate themselves.

However, if the experiment fails, the farmer usually blames the authorities. Depending on what kind of support he wants, he can portray the authorities in either a good or bad light. For example, if a farmer's experiment reveals that a certain crop variety does not meet his conditions (despite the fact that no variable has been controlled for) the farmer does not blame his experiment but points his finger at the authorities. In the case of such a failure, farmers consider the results of experiments, not limiting themselves to the conditions of their experiment but also including the political context within which the experiments took place. For example, in 2001 it was fashionable to blame any crop failure on the agro-industry and AGRITEX for siding with the interests of the big white commercial farmers and western powers in creating useless seed to discredit the government.

> The new hybrid seeds are no longer having a positive impact on our lives. If you put a little fertiliser then you get nothing at all. This year we put two bags on top and the maize became yellowish whereas with the old seeds if you put fertiliser the maize would become deep green. We know what's happening. These white men are manufacturing good-for-nothing seed so that we turn against our own government if things fail to work out.

If the experiment has been successful then farmers do not need to transfer their attention elsewhere. After confirming that the seed is good, there is no need to refer to agro-business or politics. Politics is only relevant when explaining failures.

There are those good farmers who are said to belong to the green group, or the gold group, depending on whether they are good at growing maize or cotton. Those in the green group usually get preferential treatment when seeking loans and they can also obtain a portion of free seed to experiment with, but they have to work closely with the AGRITEX officers who tell them what to do to ensure a good crop. As noted by Hedrick (1918: 162) with respect to America, and applicable to Zimbabwe today, most demonstrations are 'put on by being arranged for in advance through getting a farmer to make himself a model in performing some farm feat'. In the end other farmers are invited to observe the demonstration plots. One farmer confirmed that they learnt about new varieties by observing the demonstration plots at field days:

> That is where we usually get to know about these new varieties, because the people in the GFC (referring to green groups) are given the maize varieties for free to plant and then we go to observe the maize in their fields on field days.

Such experiments rarely fail because the farmer is supplied with enough seed and fertilisers for the portion under experimentation.

These days, according to Mr. Mushayi Mapeto, the Cottco representative, Cottco no longer provides fertilisers but just the seed. They look for farmers who will be able to afford fertilisers, and then provide the seed and supervision. However, if a farmer is able to beat other farmers and a field day is held at his or her home then that farmer will get full inputs equivalent to one hectare. If tended properly a farmer could get somewhere in the region of Z$162,000 (US$1700, using the official exchange rate of 2002) from the hectare. When asked how they had selected a farmer for a demonstration plot in Mudzinge, the AREX officer alleged that he had intentionally selected a farmer who could purchase fertiliser:

> I was looking for someone who would be able to purchase fertilisers because Cargill was only providing the seed. I first talked to the husband who said that he would be willing and able to buy the fertiliser. I met Mrs. Chenjera recently and she told me that although the seed got burned because of the prolonged dry spell we experienced, she got more maize from the demonstration plot than from any other part of her field. The problem we faced this year was that fertiliser was very difficult to get so it was quite difficult to get people for the demonstration plots.

The difference between these experiments and those carried out without outside assistance is that the success of the experiments with outside assistance is attributed to authority and to a lesser degree to the farmer. However, this farmer is considered good because of his association with the outside agencies. Failure is then attributed to uncontrollable variables such as illness and erratic rains.

The demonstration plots are very useful in the production of knowledge in the resettlement areas. This is because those farmers who cannot afford to carry out experiments on their own can have access to experimental results in the demonstration plots and decide either to adopt or not to adopt any new crops or seed. AREX also uses these demonstration plots as a way of fostering new farming ideas and practices among farmers.

> We have a demonstration plot for seven varieties of pannar seed. It is very near to this place. In that plot we also have one variety of soya beans but unfortunately birds are attacking the soya beans. We want to encourage farmers to adopt other crops like soya beans, not to just keep on planting maize, tobacco and cotton.

AREX uses the demonstration plots as a way to experiment with soya beans so that people are in a position to adopt soya beans with confidence. However, the experiments in the demonstration plots are sometimes like self-fulfilling prophecies, especially because they are not carried out in the 'true' scientific spirit of sniffing out the truth about the seed, but rather as a way of advertising certain seed varieties. This was the case with the pannar maize seed variety, which had become very unpopular with the farmers after the 1994 drought (See Chapter 5). In 2003, the Pannar Seed Company sponsored a field day as a way of re-launching its seed.

Yet, it is debatable whether these experiments in demonstration units are relevant to farmers, given the fact that most can never afford the kind of inputs that are needed for this kind of farming. For instance, all farmers in the sample complained of the cost of fertiliser, which limited its application, and now the seed is expensive, if it can be found at all. In the 2002-2003 season, a fifty kilogram bag of fertiliser was selling at Z$9,000, an increase with regard to a cost of between Z$1,500 to Z$2,000 the previous season. People said they needed three bags of ammonium nitrate per acre, and two of Compound D, and the producer price of maize increased from $Z7,500 to Z$28,000 per ton.[42] The cost of these required fertilisers then reached a price of around Z$45,000 per acre. This exludes the cost of seed, which was being sold on the black market at prohibitively high prices, as well as the cost of labour. At the time the official exchange rate to the US$ was USD1 to ZW$55, and the illegal market exchange rate was pegged at USD1 to ZWD600. Thus twelve households in the sample (four of the households belonging to the medium-wealth category) experimented with Open-Pollinated Varieties such as hickory king, which does not require much fertiliser and can do well with cattle manure. As a result, the relevance of the demonstration units to farmers was very debatable. While some farmers maintained that they had learned a thing or two by observing demonstration units at field days, others regarded the field days mostly as social events. The following two quotations indicate these two contrasting viewpoints regarding field days:

> I started to cultivate SC501 in 1995. I attended a field day, that is how I got to know about SC501. So I first experimented with the seed before I decided to grow it full scale. I planted 2 acres of R215 then an acre of SC501. At harvest time my one-acre of SC501 equalled 1 1/2 acres of R215.

Commenting on field day attendance, one farmer had this to say:

> People select where to go. For example, when people know that there is very little food they do not go to the field day. However, this is usually rare. If you do not get any food it's usually because you will have arrived late.

---

42 In the 2003-2004 season, the price of fertiliser went up to between $10,000 and $15,000 per 50 kilogram bag. The producer price of maize rose to $130,000 per ton. Thus during this season a farmer required between $50,000 and $75,000 per hectare for fertiliser alone. The official exchange rate to the US$ was USD1 to ZWD55, the illegal market exchange rate was pegged at USD1 to ZWD1,000.

Thus if the farmers do not adopt what they observe at model farms it is not because the mind of the peasant is difficult to change, as some would like to argue, but probably because the costs of doing so are prohibitively high especially if one cannot obtain loans (see also Bowden 1970 and Green and Hymer, 1966). At the same time, to some extent people are content if they can get enough for their own consumption.

Except for two farmers who carried out extensive experiments – one of them normally setting aside two hectares for experimentation, and the other planting his whole field with the experimental seed, which he will change the following year if things do not work out – most initial experiments were carried out in gardens. The difference between these two farmers and the other farmers was that the former were relatively well-off and probably could cushion losses well if the experiment resulted in a loss. However, invariably the portion under experimentation turned out to be smaller than the portion which was not. Poor farmers also carried out experiments with some Open-Pollinated Varieties (OPVs) but they regarded the OPVs as a high risk since they had no official backing. As a result women would plant the OPVs in gardens first before they would adopt them wholesale. Gardens were also used for seed breeding to ensure that there was enough seed left from a successful experiment to move the crop from the garden to the field.

> This year I planted *bharabhara* in my garden to see how it would fare and if I manage to save a lot of *bharabhara* seeds from my garden I am going to plant it in my field this year ... No one gave me the *bharabhara* seed. What happened was that Mr. Chipoira dropped a *bharabhara* cob when he was coming from his fields during harvest. I picked up the cob and planted it in my garden so that I could generate my own seeds.

On being asked if they cultivated Open-Pollinated Varieties some farmers said they did not, only to mention on later occasions that they did cultivate them in gardens on an experimental basis, for seed breeding or for consumption. Gardens are important when trying to understand experiments and the diversity of seed and crops that people cultivate. Elsewhere Howard-Borjas (1999) notes that women's home gardens are important for Plant Genetic Resource Management (PGR). She also maintains that it is in home gardens that women domesticate wild plants and adapt new crops. Hence, by focusing mostly on the fields, a huge amount of data on experiments and seed and crop varieties can be lost.

Although the soil type in the area does not vary much and farmers receive the same rainfall, two different farmers experimenting with the same seed could reach different conclusions. Below are two views of farmers on the same variety of maize:

> In 1998 we got 7 tonnes of maize. However we did not grow 401 (*Katsoko*) in the whole field. The larger area was sown with R215 and we planted only 20kg

> of SC401. We planted SC401 because we had heard from others in the village that it was a variety which matured early. After that, we never grew *katsoko* again because the maize cobs were very small.

The second farmer, Samson (not in the sample), who had first cultivated *katsoko* (SC401) as an experimental seed, was satisfied with the seed and intended to increase the portion of *katsoko* under cultivation in his field.

A key question to pose is: what exactly is an experiment? If experiments entail the controlling of variables, which variables should be controlled? Are farmers controlling any variables at all and is it possible or even desirable for them to do so? To some extent farmers do not regard these experiments as reliable. For example, if farmer A's experiment fails, farmer B does not rule out the seed variety completely. The only variable that farmers control in these experiments is fertiliser application. However, frequently it is difficult to control this because the amount of fertiliser a farmer uses depends on whether s/he has managed to get enough fertiliser either on loan or through purchase. Thus, by observing farmers' experiments, one cannot say beyond any reasonable doubt that such and such a crop or seed variety is good or bad. That is, by their very nature local experiments are specific to the individual farmer. Hence, using farmers' experiments one cannot conclude that a particular variety of crop is suitable for the area. For generalised application, we can only rely on experiments carried out by 'scientific' seed breeders. However, some farmers may disagree with the assumptions made by the seed breeders after carrying out their own experiments, when things do not work out for them as they had expected.

In some cases, necessity forces farmers to experiment, with the hope that things will improve. I am not sure whether such activities can be regarded as experimentation or should just be seen as 'a scatter bomb'. In such cases farmers acquire their knowledge on particular crops and try to apply it to other crops on an experimental basis, regardless of whether they belong to the same family or not. For example, in the agricultural season of 1985-1986, maize was attacked by army worms. People who could afford cotton chemicals used them to spray the maize crop in an attempt to kill the army worms. This was not done following the advice of AGRITEX or any other expert but people reasoned that if it could work on cotton by killing all those cotton pests, then it could also work on maize. (As it turned out in the recent army worm attack (2002) AREX did advise farmers to use a cotton chemical carbaryl 85 to spray their maize fields when suffering an attack by army worms. However, they were required to use specific doses.) Often when farmers carry out experiments, whether out of necessity or out of genuine experimental spirit, their knowledge of other crops is called upon. Thus, in the early days of resettlement, some farmers (although no one admitted to this but instead all pointed at others) had applied a portion of their cotton fields with manure just as they did with maize. They discovered later that the cotton simply grew tall and green but that no balls formed, so they stopped doing this. However, the AREX officer had a different opinion on the use of manure on cotton:

> If you use manure on cotton, the cotton will have a very high rate of growth. The cotton plant will grow very tall. What you should do if you want to use manure on cotton is first apply manure to a maize field, then rotate maize and cotton the following year. For cotton, you will also have to use compound L or soluble boron.

Animals

Experimentation is not only limited to crop farming but nowadays has also expanded to animal rearing. In the early years of resettlement, farmers worked very closely with the veterinary officers on preventing and curing animal diseases. However, because of the escalating costs of medicines to treat animal diseases people became more experimental. Experimentation, like other activities, is gendered. In this case, it is not that men experiment more than women or the other way round, but that men and women usually experiment with different things. Crop experimentation and animal experimentation are different. It is mostly women who carry out experiments with crop varieties in their small garden plots and breed seed which can then be used in the field if successful. And it is usually men who experiment with animals.[43]

When it comes to experiments regarding animal rearing it is mostly men who deal with the larger livestock such as cattle and goats whilst women can sometimes deal with small livestock like chicken. Experimenting with crops is usually done to find better varieties, which can suit the demands of each farmer; however, experimenting with animals is usually carried out to find cures for diseases, not to find better breeds.

The reason why experimenting to find cures for animal diseases is mostly the prerogative of men when larger animals are concerned, is that the big domestic animals like cattle have been 'masculinised'. Women cannot keep cattle where they are married. If they have cattle they have to be kept at their fathers' or brothers' houses. One old man maintained that it was a cultural taboo for the wife's cattle to be kept in the same kraal with her husband's cattle. If such an anomaly were to occur the wife's spirits would become strong and kill all her husband's cattle. Although sometimes women refer to the family herd as 'ours' there was an acceptance by women that the cattle are theirs only on a usufruct basis. This is reflected in disputes that some-

[43] Elsewhere, in Malawi, Ferguson (1994: 543) noted that 'knowledge about crops and responsibilities for growing them is often differentially distributed by gender'.

times arise between husband and wife over the use of money gained from farming:

> Sometimes disagreements arise. For example, this year I felt my needs were being ignored. Every year we were buying cattle and these belong to the men. We had never bought anything I could call my own. As a result I felt it necessary that I should get something. My husband thought otherwise. We had a huge misunderstanding; other people in the village had to intervene to solve the dispute.

As a result, women do not focus at all on cattle because they do not own them and if such an animal experiment was to fail it would back fire on them. One very poor villager lost the only ox he had because he was away and his wives did not want to try a traditional method they had observed to cure eye cataracts. Most people we talked to on this issue agreed that if the wives had used the method the ox would have been saved. When asked why they had not tried to save the ox, one of the wives had this to say:

> The problem was that father (i.e. the husband) was not there. He came back after the ox had died. We mourned for that ox as if we were mourning for a dead person.

However, widows to some extent experiment with cures but first they ask advice from some knowledgeable man in another village where they do not have grown-up sons.

Experiments with large animals are rarely carried out on an individual basis. When an experimental cure is being tried friends usually carry it out together so that everyone can observe whether the animal is cured or not. Just like people sometimes refer to witchcraft as science, so experimenting with cures for animal diseases is described in this way, though crop experiments are never. In this case, the term 'science' seems to refer to the mysterious workings of witches and antidotes used to cure animal diseases that are not easily understood and explained, as compared to the relatively straightforward nature of crop farming. One example of such 'science' concerns how a group of men decided to mix traditional beer with cooking oil and give it to calves to cure *gwembe* (a skin disease in cattle):

> You mix the scud[44] with some cooking oil and give it to the calf to drink. I think it is the bitterness in the scud that kills the disease. It's like our traditional

---

44 'The scud' refers to opaque beer manufactured by Chibuku breweries. The opaque beer is also known as Chibuku beer. At the height of the first gulf crisis Chibuku repacked its beer into new containers. People said the new containers of the opaque beer looked like Scud missiles henceforth the beer and its container became known as the scud.

medicines, the more bitter they are the more effective they are. Even those malaria tablets are bitter. We just decided to try it to look for 'science' to cure our animals.

But a scud, how can someone think of that?
What happens is, when we are with other men, one person can just start to muse aloud on whether a scud can cure diseases. After some discussion, someone with an ill calf can just offer his animal for the trial. That is how it works.

This is how knowledge is manufactured not by rigorous laboratory experiments but by a group of men who just happen to think that traditional medicines work because they are bitter, just like chloroquine is bitter and works against malaria. The thinking does not end there but continues: because of its bitterness the traditional brew, the scud, must be good for something apart from making people drunk, so let's try it on animals. Of course had a scientist been present when such a conversation was taking place he would have quickly pointed out that neither chloroquine nor traditional medicines work because they are bitter, but instead because of certain chemical components that they have. But because no such saboteur is there the discussion moves to the second stage, concerning who is willing to sacrifice a calf for the experiment. After the person is found, and the experiment has worked, soon everyone in the village becomes a 'veterinary officer', making scud and cooking oil concoctions to cure the various illnesses of their livestock. Now the cure is not limited to cattle but is also tried on goats and chickens. Everyone believes that it is the bitterness that works. At this point perhaps, somebody comes up with the idea that maybe it has something to do with fermentation. In the meantime, no one has been able to explain exactly why one needs to mix beer with cooking oil, except in cases where cooking oil is used to assist with bowel movement when a goat has over-eaten.

After the initial experiments are carried out, knowledge becomes something that people have and use but they may not know how they came to know what they know. People will also not question themselves as to why they do what they do, as long as it works.

Experimentation with animals, as I have said, is not limited to men; women also carry out experiments albeit limited to smaller animals. They also try unorthodox remedies, such as giving their chicken paraffin to cure coughs or smearing the whole chicken with paraffin to get rid of pests. The main argument here is that the smell of paraffin chases away pests and because it is poisonous aphids will die. As to why the chickens do not die after having been fed with the paraffin, it is said that this is because the paraffin is given in small quantities. In addition, the chickens do not swallow the paraffin because the taste is not pleasant and in trying to force the paraffin out of their mouth they clear their throats of the cough. Thus, taste and smell, and not chemicals are the main components in search for a cure, unless the potential medicine is considered to be highly toxic even when given in small doses.

People experiment more and more with their animals mostly because the costs of medicine have increased prohibitively.

> We only go there for serious diseases that we do not know. If you go to the veterinary when your cow is suffering from *gwembe* you will end up buying medicines worth over $2,000 ($36) instead of buying a scud which costs less than $150 (US$3). Our own ways are much better. If the disease does not get cured and you are worried then you can go to the vet ... In the past people were using those chemicals from the vet because they were cheaper. Now people have decided that going to the vet is a waste of money when the animals can be treated cheaply.

This lends credence to the saying 'necessity is the mother of invention'. Linked to this, there is a general mistrust of veterinary officers because some people suspect them of advising people to buy very expensive medicines so that they can safeguard their jobs, whereas cheaper medicines, which are equally effective, are available. In both villages, people only inform the veterinary officers if there is an outbreak of serious diseases such as anthrax where the law requires that they notify the relevant authorities; and when, if the necessary preventative measures are not taken, the whole herd of cattle could be wiped out.

The above data should not suggest in any way that people are now more experimental than they were in the past, since there are no reliable data for comparison. Some diseases that farmers encountered after resettlement were new to them. Before resettlement, farmers kept disease-resistant and drought-tolerant 'hard *mashona*' cattle. However, after resettlement, they started to keep new breeds of cattle. These latter were also attacked by strange new diseases often referred to as 'diseases of modernity' (*zvirwere zvechirungu*). Farmers would experiment with old remedies, or look for alternatives if these did not work. Local knowledge does not die because of exposure to outside knowledge. Instead people actively select those aspects of both knowledge systems and adapt them according to the demands of the situation. Most farmers were adamant that the only cattle diseases they were aware of before coming to the resettlement schemes were liver flukes, diarrhoea (which was caused by changing seasons) and eye cataracts which could easily be cured by the use of *mutamba* fruit or fermented *rapoko* (finger millet) flour. Farmers agreed that with cattle diseases such as black leg one did have to consult the veterinarian.

However, in 2002 there was an outbreak of black leg in Muringamombe and, unlike other villagers, one poor farmer decided not to use the expensive modern preventative medicine available through group purchase. Instead he applied aloe vera, a known folk medicine, and none of his cattle were affected by the disease. On an earlier occasion, this same farmer had mentioned black leg as the only cattle disease that would cause him to consult the veterinary services. Insofar as he was using the aloe vera against black leg for the first time, his application can be said to have been experimental since he did not know whether or not it would work. Apart from using aloe vera, this farmer said he made sure his cattle grazed at the river where the grass contained water. He maintained that it was the grass the cattle were grazing that caused black leg, because the grass had not been rained on for a long time and hence it was full of disease.[45]

> *But you told me earlier that black leg is caused by mud.*
> That is during the rainy season but now it is because of the grass. The grass is weakening the cattle's bones. That is why the cattle are being attacked in the legs.

Diagnosis and experimentation are not only the privilege of the rich farmers. Although the above respondent was poor he used his small number of cattle as an excuse for not using modern medicine when, in actual fact, it was because he was not on friendly terms with the people who had bought the medicines and who were distributing them to most of the villagers. A few weeks before, the wife of this respondent had been accused of witchcraft at a village meeting. He denied that his wife was a witch and maintained that this was because the other villagers hated him.

Observation

Crops

Observation is a central process in learning about farming. Observation can be covert or overt. Although most people in the sample did not admit to learning by observation, it could be noted that some of the answers they gave pointed to this effect. In both villages, expressing undue interest in your neighbour's field could trigger an accusation of witchcraft; as a result people did not mention observation as a component in their acquisition of knowledge. It was also not only the person who was observing the field who could be accused of witchcraft. In at least two cases, respondents mentioned that they would not go to some people's fields to observe because these people had bad medicines. Strange things would happen to those who dared to observe fields without permission from the owner:

> You might hear strange noises. Sometimes you will hear voices but you do not see the people. You can be instructed by a voice to leave the field immediately!

---

[45] According to the AREX officer, bacteria cause black leg and penicillin is the only cure.

> Even snakes (*he starts laughing*) ... if you go into these fields without permission you can be chased away by very large snakes.

Some people with farming magic were known to be hostile to those who wanted to observe their fields because these observers would not know of all the taboos required. If such taboos were not followed then this could spell disaster for the owner of the field. For example, marauding animals might attack his crops, or the crops would simply refuse to do well.

Despite this denial of learning through observing other people's fields, observation is still central. One woman was asked how she had come to know of herb killers, although she had never attended agricultural lessons, did not go to field days and did not observe other people's fields (for fear of being labelled a witch). She answered that she had observed that in a field they passed on their way to church the field no longer had a problem with weeds. As a result she asked the owner of the field, whom she knew on a personal basis, what they had done with the weed and she was told that they had used a herbicide.

What is interesting to note is that, although they referred to observing other people's fields, most people mentioned observing the fields of others who were living in a different village. People were only expected to observe the fields of friends. If the person was not their friend, then observation would only be possible if that person hosted a field day. One person was threatened with a beating after he passed through the field of a farmer he did not know in another village with the intention of observing. The owner of the field suspected that his passing through his field had something to do with the use of bad *muti*.

Only one person admitted that he went to observe the fields of his friends who lived within the village, and that sometimes they would compete with each other over who was going to achieve the best crop. On another occasion, the same person mentioned that going to people's fields to observe was a thing of the past. On being asked why this was so, he said:

> In the past people were united. Now! Ha! Even the AGRITEX officer commented on it one day that in the past people used to farm and buy useful things like ploughs but now people are farming and investing in goblins (*zvidhoma*) instead. This area was developing but now everything is going down hill. In the past people here were very good farmers. A bale of cotton would sell at $5 but that money would be invested wisely for development.

This he said after the person he had mentioned earlier as the friend where he went to observe, had not invited him to a tombstone laying ceremony. Thus, in this case, the issue of witchcraft was used to cover up socially strained relationships. However, it is undeniable that observation is an important aspect of the production of knowledge in the research area, but it is difficult to point at it if direct questions are asked.

Observation is also important where youth are concerned. Although people often did not know why they performed certain tasks in certain ways, they maintained that they did those things in the same way that they had observed their parents do them. One respondent said that she had learnt how to cultivate *rapoko* from her parents.

> Mostly I learnt about *rapoko* when I stayed with my parents. *Rapoko* is something that is in people. It's not something that you learn to do or something that you can even remember learning. You just know that you know. We just learned by seeing what our parents were doing. They were transplanting the *rapoko* and doing a lot of things to the *rapoko* and all these things we saw with our own eyes.

This respondent summarised in a neat way how some things are 'in people'. Especially when knowledge has been gained through observation, it becomes internalised. Such knowledge is also generally regarded as more valuable since it is morally enforceable. Some people confessed to doing certain things because that is what they observed their parents do and their parents had also learned from their parents. (The subject of youth and knowledge will be discussed in more detail in another chapter.)

Observation can also be used for verification purposes. People can discuss new things with their friends and relatives but they can not just adopt any new thing before they are sure that it works. Even among close friends and relatives there is an element of mistrust, so that people only feel comfortable in adopting something when they are sure it will work.

To be sure, they have to observe to ascertain whether whatever is being claimed works. As a result, most people said they would only adopt a new thing after they had observed that the thing worked for the person who first discussed the idea. Below is an extract from a conversation involving the researcher, research assistant and one of the respondents.

> *(Christine (the research assistant)) Say, you hear of or see a new crop, what do you do before you decide to adopt or not to adopt the crop?*
> Maybe I will buy the seed in small quantities then plough a small piece during that year and see how it will do.
>
> *What if you hear about the new crop from your neighbour; do you just accept the new crop?*
> I look at whether the person is also growing that crop and I will ask that person about the crop. If they are not cultivating it I will ask them why they are not cultivating it, why they have given that crop up. I will ask them why if they have stopped cultivating the crop they are encouraging me to cultivate.

(Netsayi) What if it is your friend whom you trust who has come up with that idea?
Well, I will still have to make sure that they are still cultivating the crop. If they say, 'last year I made a profit', if it is true, then maybe I will also adopt the crop. One cannot accept seeds from someone just because that someone is your friend. Some people can just lie to you about the benefits of cultivating a certain crop when they have not grown the crop themselves.

Why do you say they lie? Has someone ever lied to you?
No.

So why do you say people lie? I talked to some people and they said that they feared to take ideas from other people because people lie, but when I ask them whether they have ever been lied to, they say no, they have never been lied to.
Some people lie. Sometimes in your mind you just suspect someone of lying to you, even though you have no evidence that you have been lied to before.

Although sometimes observation can breed suspicion, it can also be a very reliable method of verifying information. As shown in the above case, in a situation where there is a high level of mistrust, observation as a way of getting new knowledge or even disseminating new knowledge is indispensable.

Although the percentage of people who stated that they got their knowledge at field days has declined from 79 per cent to 29 per cent (from the 1980s to 1990s) field day/ demonstration plots still rank highly as a place where people manufacture and disseminate knowledge. Information relating to new seed varieties is passed on from seed companies to AGRITEX, whose field officers in turn disseminate the information to the farmers by word of mouth and through field days. Informed of the basic features or characteristics expected of a crop grown from that seed, AGRITEX carries out 'field observation trials'. In Mupfurudzi, the more successful farmers, especially those in the green group for maize and those in the gold group for cotton, would usually volunteer for field trials. These people would use their inputs and work in conjunction with AGRITEX so that, if their crops did very well, people would come to observe their fields.

AGRITEX and the Seed Houses help with the provision of food for field days, otherwise the farmer would only get prestige and sometimes might win ploughs and other things.

> Field days are carried out at fields. The owner of the field is the one that explains everything to the audience. The farmer will provide beer and some farmers would slaughter cows or goats. AGRITEX officials also contribute but usually they only buy a little food to help the farmer. However, the farmer could sometimes win big things like a plough.

Officers disseminate knowledge about crops, especially maize, cotton and recently tobacco, on field days. The field days give people the opportunity to observe in a relaxed atmosphere, since the occasion combines observation, learning and entertainment. What I find noble about these demonstration units is the fact that the farmer whose field is being observed as the model will explain to other farmers how he managed to produce such a model crop. Only a farmer who follows 'modern' ways of farming would have the honour of having a field day held at his or her field. This could be one of the many reasons why so many farmers have begun to adopt the new methods of farming and new technology because this new technology is associated with prestige, and of course with improved yields.

The concept of these demonstration plots is not entirely a new phenomenon. In the 1920s there was a conflict between two groups of white agricultural experts, in which one group argued for the establishment of model farms and the other for the training of native agricultural demonstrators. Here it was assumed that to improve farming the Africans had to be exposed to model farms so that they could copy the methods. On the other hand, others wanted training for native demonstrators so that they could assist fellow natives with agricultural knowledge. Howman (in Steele 1972: 13, 16) maintained that African farming was deplorable and lacked imagination and, above all, the mind of the African peasant was difficult to change. When asked about the 1920s controversy, Howman (in *ibid.*: 32) said:

> I wasn't aware of the people that advocated the model farms. At that time people took a simple attitude towards how you could change an African. They believed if you show him something better he is bound to pick it up, a fallacy, which I do not think now would be accepted anywhere. Every European farm is a model farm but it made not the slightest difference to these people.

As can be seen, the European farm was regarded as the model, so Africans underwent a process of being labelled 'ignorant', and their knowledge was delegitimised. For example, in the early days, people were forbidden to intercrop because it was thought that crops would disturb each other. After a couple of years there was a shift again towards intercropping, as it was shown that it increased ground cover and reduced soil erosion. Nowadays, AGRITEX encourages intercropping as beneficial for crops and situations of land scarcity.

Although demonstration units are used to disseminate knowledge to farmers, they also have a highly political component. When people attend field days to observe these demonstration plots people also perform dramas and sing songs of the revolutionary movement as well as praise songs for President Mugabe and his government. Hence, knowledge of hybrid seeds is disseminated at the same time as the political ideologies of the ruling party are reinforced. To be successful, AGRITEX (AREX) officers (especially in the current scenario of political uncertainty and unrest) need to be seen, at least publicly, to be aligned to the ruling party and its standpoint.

Learning by observation is very useful for those who are not very well educated and cannot attend the conventional agricultural lessons offered by AGRITEX where people have to take notes. Although people can take an oral exam for the Master Farmer certificate they do not want the embarrassment involved in being the only illiterate among literate persons. As one illiterate farmer pointed out, he had attended the AGRITEX lessons in the early days until he realised that he was not learning anything new since he had learned everything at the commercial farms where he used to work. Most of the older farmers preferred to learn by observation and did not trust any theoretical knowledge that was be imparted to them without a practical backing:

> I do not listen to the radio or even to the AGRITEX officers. Those people do not know a lot of things. I stayed a long time at the large-scale farms working so that is where l learned how to farm. I learned from the white man himself (*ndichiona mubhunu*). I learned by observing. Cotton and maize l learned from white man's farms.

In 2001, not many farmers admitted to have learned anything from the large-scale white commercial farms where they had lived and worked for several years prior to resettlement. Only one farmer mentioned that this was where he had obtained most of his knowledge on farming, while in 2002 this number had risen to five. This difference was probably the result of the tense political situation that was prevalent in 2001 during the parliamentary elections. At that time white farmers were vilified in political rhetoric as oppressors and as a result it was politically incorrect for one to admit that one had learnt anything from them.

Animals

As underlined earlier, observation is a very central element in learning how to do anything. Especially when cattle are being treated or given preventative chemicals, men and boys in the village congregate at the cattle kraals to observe the whole proceedings. Sometimes even small boys, who are too young to be of help, are invited to observe so that if they ever face that same problem in the future, they will recall the knowledge of their fathers. The picture at the start of this chapter where people are congregated at a kraal where cattle are being vaccinated against black leg disease depicts this well. Even young men whose parents did not own cattle or did not have the money to buy the chemicals turned up to observe the event. When asked why, they said they had come because they wanted to learn what to do when faced with the same problem in the future.

Young boys who were too young to take part in the physically strenuous activities were given small tasks to do. Most people learn in this way to perform certain tasks at an early age, and often they cannot pinpoint the exact moment this happened. A more common response was, 'no one told us, we just know'. On being asked how they treat eye cataracts in cattle one male respondent explained:

> We use traditional methods. We take an unripe *damba* mix it with a little water then drop the liquid into the eye of the cow/ox. You can also use *rukweza* (fermented finger millet flour). You put the *rukweza* in the eye of the animal.
>
> *Do your young children know all these things you are telling me? (The two young boys start to laugh.)*
> They know everything because they also help to treat the animals. For example, if a cow or ox gets *tsanga* (eye cataracts) you can ask the young boys to look for the *damba*, which will be used for the treatment.

Knowledge is practical. When smaller livestock such as chicken are given doses of paraffin to cure diseases, even young children will often be told to chase and catch the chickens and therefore will be present when the treatments are administered. Young children, therefore, are not only observers or consumers but they also play some minor productive roles that prepare them for their adult lives. As Keesing (1987) puts it, knowledge becomes like the layers of an onion, where one keeps peeling until one reaches the core. Knowledge is not imparted all at once, but in bits and pieces, as people progress from childhood to adulthood, so that there is never an exact turning point that a person can mention as the moment he/she obtained this knowledge. Where observation is concerned, especially when learning in a family setting, knowledge can only be said to be incremental.

In Zimbabwe there are not many taboos associated with animal rearing. When people claim that their enemies have killed their animals they do not imply this was done by mystical means. It is usually by poisoning, drowning or axing the animal to death. Usually this is done in retaliation, for example, after a dog has mauled the other person's animals or a cow or ox has grazed in another's field. Sometimes it is for a religious reason such as when people accused Marange Apostles of poisoning dogs in the village because the apostles believe dogs are unclean in the eyes of God. As a result, it is easy to learn by observation where animal health is concerned, since anyone can just observe animals being treated without being suspected of magical malicious intentions towards the animals.

I am not quite sure why there are so few taboos regarding animal rearing. This might be related to the fact that when cattle are ill their illness is easily diagnosed and actions can be taken to remedy the illness. On the other hand, when it comes to crop farming, people cannot control many of the critical variables including soil fertility. Faced with the many variables they cannot control or understand, people are perhaps more inclined to believe in magical explanations.

Conclusion

Both farmers and scientists carry out experiments, although they reach very different, and even sometimes conflicting, conclusions. Although these experiments differ in their construction and basic structure, both farmers and scientists are out to improve livestock and crop farming. Since they consider different things when evaluating seed, if the scientists fail to consider the needs of farmers, they can breed a seed of high scientific quality but one that does not meet the needs and demands of farmers. There is thus no rigid distinction between 'nature' and 'society', or between 'culture' and 'science'; they are embedded into each other, as this chapter has demonstrated. This proves the fallacy of the modernisation theories where knowledge is regarded as only that which flows from the experts to the farmers. Farmers should not be viewed as traditional and unscientific, but as actors who actively take part in the production of knowledge by making situated selections of what they think will work for them, and who also experiment to improve agriculture in a way that is meaningful for them. This holds true regardless of whether what is meaningful for them conflicts with 'expert' knowledge. Knowledge and indeed the capacity to produce more knowledge is not only the preserve of experts. As a result, there has to be at least some form of consultation between experts and farmers before seed can be developed.

Secondly, although the sponsored experiments were an essential way for farmers to gain knowledge, they were of a limited usefulness because of the high input cost required. Sponsored experiments were good for farmers who could afford to buy seed and fertiliser, for they could purchase seed with full knowledge of its potential, thereby minimising the risk. As individual choices were limited by the distribution of resources, it is my conclusion, however, that for demonstration plots to be relevant, agricultural experts also have to focus on low-input farming with better returns. These experiments could focus on the use of different kinds of manure and locally available seed. These experiments would be highly relevant to the needs of the poor, ensuring that all social levels of people are catered for. It should be noted that, even in the event of a demonstration of an agricultural 'feat' performed by some farmer, farmers can only take up practices that they can afford. Thus farmers do not always stick to what they know because they are conservative, but often because the conditions under which they operate set limitations to what is possible. Farmers have to do the best they can within limited choices. For farmers it makes much sense to observe the fields of friends and adapt their friend's ways of doing things because usually they experience the same structural circumstances, instead of observing demonstration plots that emphasise high-input agriculture when they can not afford this. In this way, individual choices are shaped by the distribution of resources.

As people are generally more sceptical of expert knowledge and rely more on their own observations and experiments, AREX and other knowledge experts must tap into this proclivity. As discussed in this chapter, for most farmers observation is important before any new methods of farming are adopted. There should be more observation plots. However, these demonstration or observation plots should be structured in such a way that they address problems as defined by farmers, not as defined by experts. It should be realised that expert knowledge is mediated by farmers' experiences, beliefs, and ability. Farmers make selections and choose that which suits their own needs and capabilities. Although there is a need to equip farmers with theoretical knowledge, Master Farmer certificates need to be regarded as more than just paper to access resources.

To be effective, AREX and other experts should understand people's belief systems. Frequently, people's beliefs can prevent people from adopting - or sometimes encourage them to adopt - certain behaviours. It is important then for rural development workers to understand the belief systems of the people they work with, since this will enable them to negotiate the beliefs and changes in behaviour without antagonising people.

This chapter has also challenged the notion of the 'conservative' peasant farmer. Such a notion does not provide a convincing picture since local farmers are not homogeneous in their outlook and neither are they very conservative. This fallacy of labelling peasant farmers as conservative stems from another fallacy, namely, the modernist belief that views 'development in terms of a progressive movement towards technologically and institutionally more complex integrated forms of modern society'. This is a fallacy because, as people are confronted with modernity and its problems, they find local solutions that are more beneficial for them than modern remedies. Farmers make strategic accommodations whereby expert and local knowledge are integrated in novel ways. Thus sometimes farmers do not refuse modern methods because they are insufficiently tutored or because they are not yet believers in science and need to be converted, but rather because they choose, in certain circumstances, to strategically break away from official knowledge.

It has become apparent in this chapter that farmers also have a thirst for producing new knowledge. They do not learn from only doing what they have always done, but actively experiment to produce knowledge.

5

Magic, witchcraft, religion and knowledge

Introduction

There is hardly a way to talk about farming and knowledge in Mupfurudzi without reference to issues of magic, religion and witchcraft. Indeed, these issues had a way of cropping up in conversations with various informants whether they were specifically brought up or not. People would deny or acknowledge the existence of magic depending on the social context in which a particular question on magic was asked. However, no one in the sample denied the existence of witchcraft and all respondents claimed to belong to one religion or the other, Christianity or African Religion.

Letts (1991: 305, 306) regards magical beliefs, religious beliefs, and superstition as illogical, inconsistent and evidentially unfounded. For him the only legitimate questions that can be asked of these beliefs are why people hold them, where and when they originated, how they are transmitted and what functions they serve. For positivists like Letts, there is a real world out there in which people need to act objectively and scientifically to achieve results. To them, believing in witchcraft or magic is just like believing in Father Christmas and the tooth fairy. In this view, to be effective, local farmers should react to the dictates of the objective world and their magical and religious beliefs are considered as being out of touch with the 'real' world out there. Magical, witchcraft, and religious beliefs are regarded as retrogressive and as an obstacle to change. For Bourdillon (1989: 29) the term 'magic' can also be used to denote circumstances where 'people confuse the logic of communication with the logic of material efficacy'. His argument is that this kind of confusion sometimes occurs and that it is convenient to classify such confusion as magic.

Accusations of witchcraft have been linked to jealousy (Fisiy and Geschiere 1996: 197; Daneel 1971: 68; Dolan 2002: 669; Ciekawy and Geschiere 1998: 5), and for Evans-Pritchard (1937: 404): 'The sickness is the sorcery and proof of it'.[46] This denies the possibility of regarding people who believe in witchcraft as rational people who consider evidence before witchcraft is attributed to be the cause of anything. Instead the witchcraft accuser is condemned to a perpetual state of jealousy and confusion in which he fails to recognise the all-too-obvious link between disease and bacteria and other contaminating agencies out there in the 'real' world.

Although I do not deny that some witchcraft accusations are a result of jealousy, in most cases people consider the evidence before them before accusations are made. Some witchcraft accusations are dismissed by villagers (both rich and poor) for their lack of evidence. Although Niehaus *et al.* (2001: 116) recognise that often evidence is needed to ascertain whether witchcraft has occurred, he trivialises this evidence when he claims that sometimes evidence can be circumstantial. Thus, if his view on circumstantial evidence is taken to its logical conclusion, evidence that is normally permissible for those who believe in witchcraft is not substantive, is not able to link the witch to the witchcraft act, and indeed does not even prove that witchcraft has occurred at all. Where the witches confess, the confession is tied to political power games. Similarly, the poor are seen as using the confessions or threats of witchcraft to gain power within their households (see Dolan 2002: 667, on how women in a district in Kenya used threats of witchcraft against their husbands to gain access to resources within households). Niehaus *et al.* (2001: 9) regard witchcraft as a 'weapon of the weak', which they use to gain access to resources owned by the rich within their communities or families.

Elsewhere witchcraft beliefs and accusations have been linked to conflict and stressful situations. 'In our extremely stressed society, traditional witchcraft beliefs provide apparent relief. At times of economic repression, suspicions of witchcraft abound as do the consequent witch hunts' (Bourdillon 1993: 119; see also Dolan 2002: 663 on the link between witchcraft and friction within communities). Although these observations are relevant, there is a need to go beyond these economic and social tension approaches to witchcraft. This is so because as mentioned above, people consider different kinds of evidence before a person is accused of witchcraft, in spite of his or her wealth or lack of it.

---

[46] However, it should be noted that Evans-Pritchard (1937) tried to show that witchcraft beliefs are rational and based on experience although he also was interested in explaining why the Azande people believed in these false beliefs. For instance in explaining why the Azande believed in magic in spite of contradictory evidence and beliefs, Evans-Pritchard (*ibid.*: 475) wrote: 'Magic is very largely employed against mystical powers, witchcraft and sorcery. Since its actions transcend experience it cannot be easily contradicted by experience ... contradictions between the beliefs are not noticed by the Azande, because beliefs are not all present at the same time but function in different situations. They are therefore not brought into opposition'.

Structural approaches that emphasise economic explanations and the accompanying jealousy to some extent fail to analyse magic and witchcraft beliefs from the point of view of the people who believe in them, or who believe that they are witchcraft practitioners. This is so because these people's beliefs and their confessions are dismissed *a priori* as due to delusions caused by confusion and a critical failure to deal adequately with 'objective' facts in an 'objective' world; as an ineffectual attempt to deal with modernity; or simply as an indicator of the stress levels within a society. It will become apparent in this chapter, that instead of adopting structural explanations, witchcraft beliefs are better understood by adopting situational and multi-meaning explanations which are highly aware of the dynamics and differential interpretations of the significance of witchcraft.

The witchcraft-modernity thesis that links beliefs and accusations of witchcraft to an attempt to deal with the malcontents of modernity (Geschiere 1997) fails to explain why people who believe in witchcraft at one moment might not believe in it at the next moment, except maybe by simply alluding to jealousy and the need for power and wealth. If witchcraft beliefs are to be understood as an attempt to deal with the 'malcontents of modernity', what does it mean in circumstances where the believer in witchcraft does not turn to witchcraft explanations, when in other similar circumstances s/he does? Does it mean that, in that particular instance when s/he chooses not to resort to witchcraft explanations, the tension between him/her and modernity's malcontents has been resolved? This then leads to a spurious understanding of modernity and witchcraft whereby if people do not understand or fully appreciate or want to gain control over modern changes, they turn to witchcraft, whereas, when for some reason or other, understanding and enlightenment finally dawn on them they drop witchcraft accusations and beliefs until a next time when, again, they are in tension with another segment of modernity. Such understandings can only be achieved through denying persons the capacity to interpret the evidence available to them, and the right to be able to believe or not to believe, and they only focus on the possibilities afforded by these discourses to gain control over modern changes. This chapter maintains the position that there is a need for 'anthropologists to avoid either extreme in their analysis of witchcraft: on the one hand of assigning only the 'traditional', micro-community social tensions to witch-like powers, or on the other hand of superimposing western academic and popular notions of modernity onto them' (Rassmussen 2004: 336).

Geschiere (1997) laments what he calls 'the rise of the occult' in modern Cameroon and links this to modernity where politicians are seen to amass power through a recourse to the occult, where witchcraft is a 'language that "signifies" the modern changes: ... it promises unheard of chances to enrich oneself' (*ibid.*: 24). However, what he does not appreciate is the fact that what he perceives to be a rise in the occult might simply reflect an opening up of political space in which these beliefs can be discussed openly by the repealing of the colonial laws that made it illegal to accuse others of witchcraft and seek redress in courts for any loss and pain suffered through the witchcraft perpetrated by others. The same criticism holds for Niehaus *et al.* (2001) in that, although in South Africa the Witchcraft Suppression Act has not been repealed, the fall of the apartheid regime opened up spaces for people to discuss witchcraft. Hence the apparent rise in witchcraft which Niehaus *et al.* link to modernity can best be explained as a grassroots response to democratisation, which opened up discussion on matters that had been expunged from public discourse.

In Zimbabwe, to the dismay of the colonial authorities who were intent on 'improving' African agriculture, Africans always linked good crop yield to magic, witchcraft and religion instead of to good farming practices (Bolding 2004), and skills such as hunting were, and still are, linked to magic and religion. Thus, in this chapter, I take the view that witchcraft is not a language that signifies modern changes, but rather continuity, with 'customary' and deeply embedded sets of beliefs and practices. In African societies wealth, health, and agricultural production, or rather wealth, health, agricultural production, and fertility were inextricably linked to issues of magic, witchcraft, and religion, leading to a rearrangement of witchcraft beliefs and arguments whenever the latter were contested, negotiated, and reworked by people in accordance with their needs. This, in itself, is not problematic as long as the dynamism of witchcraft beliefs is recognised and we do away with attempts to understand witchcraft simply as 'traditional' or as 'modern'.

> More often bits and pieces of discursive texts are brought together in innovative ways or in strange combinations in particular situations in order to negotiate or contest certain shifting points of view. Indeed the multiplicity and fragmentation of discourses (and this would include discourses on witchcraft) ... is more often the case than the clash of well defined opposing view points and rationalities' (Long 2004: 28).

For Geertz (1966: 4, cited in Keesing 1987: 166),

> ... religion is a system of symbols which acts to establish powerful, pervasive and long lasting moods and motivations in men by formulating conceptions of a general order of existence and clothing these conceptions with such an aura of factuality that the moods and motivations seem uniquely realistic.

A weakness in Geertz's definition is that he does not leave room for negotiation, contestation and change in his definition of religion. In this chapter, there is a realisation that there is not one religious belief, and that religious beliefs in the area I studied were contested (as indeed they are contested all over the world). This chapter acknowledges that different people adhering to the same religious faith might not practise and interpret their commitment in the same way. However, regardless of these differences within and across religions (both Shona religion and Christianity[47]), religious belief affects conceptions of knowledge and how agriculture is practised. As Havekort *et al.* (2003: 142) put it,

> ... religious and philosophical concepts have their place within traditional world views. Cosmovision to a larger extent dictates the way the land, water, plants, and animals are to be used, how decisions are taken, problems are solved, experimentation takes place, and how rural people organise themselves.

However, people do not simply order their practices according to a pre-given order of religiosity, but, as will be demonstrated with respect to *chisi*, they invent new religious forms and transform the old religion to suit new needs and conditions.

On the other hand, there are other scholars, like Pool (1994) and Jackson (1989), who do not seek to establish the truth or falsity of these beliefs, but rather to study and understand these beliefs in their contexts. Why spoil the fantasy? In this chapter, however, I do not aim to discuss the various functions or dysfunctions of various magical, religious and witchcraft beliefs that people may hold. Neither do I wish to discuss the premises on which these beliefs are held to be true. Denying the importance of these factors, for example in enhancing or lessening agricultural production as scientists do, does not make them any less important in people's agricultural practices. My aim in this chapter is primarily to discuss how these beliefs impact on knowledge. The importance of magic, witchcraft and religion as part of the local theoretical tradition for identifying facts and genuine phenomena can hardly be overemphasised.

Magic

While much agricultural practice is shaped by economic and technical choices, there are also issues of culture that affect people's perceptions and behaviour. In this section, I focus on magical beliefs and practices that can affect farming practices, and in particular that can affect the way information about agricultural technology is produced, transmitted and received. I will

---

47 As noted later in the chapter, some independent Christian churches have borrowed heavily from African Religion. In spite of this, these churches have managed to maintain their distinction from African Religion, just as they have managed to maintain their independence from mainstream Christian churches such as the Roman Catholic Church, the Methodist Church, the Anglican Church, the Salvation Army, etc.

try to distinguish between religion and magic, but they feed into each other and cannot always be distinguished. Some forms of magic are incorporated into religious practice but not all religious experiences are magical and not all magical experiences are religious in character. For example, some people regard water blessed by Christian priests and prophets as an effective counter against people with bad magic. On the other hand, those who believe in African religion might interpret events such as poor yields as a punishment from the ancestors who have great power over the livelihoods of their descendants. Indeed for me, religious beliefs are magical. My decision to separate magic and religion is based on the fact that most Christian churches regard magic as evil and they ban the use of magical amulets and medicines among their followers. Thus, in the Johanne Masowe and Johanne Marange churches anyone who is seen using these magical amulets or magic horns is regarded as a witch.

These beliefs in religion and magic provide a basis through which people evaluate their own performance as well as the performance of others within the community. For instance, good crop yields might mean a person is using bad magic, or that s/he has protected his/her field with magical charms, or that the ancestors and God are satisfied with his/her conduct, not merely that s/he has great farming knowledge or that s/he manages his things well. For many, the consideration of effort or skill or technology is a secondary issue – and in any case poses the question of how an individual acquires the skill or knowledge or propensity to work. Often magic is supposed to produce good yields at the expense of others: for instance, the magician finds supernatural ways of stealing the crops of others. On the other hand, when good fortune is understood in terms of religion, then it can never be at the expense of others, and sometimes it will be good for the general populace.

The majority of the people in the sample believed in the existence of magic to enhance agricultural skills. However, of all those who believed in the existence of magic, none admitted to having used magic and only two people in 2001 and three in 2002 claimed to have lost their produce to people with magic.

<u>Case 1</u>

In this family, the children were convinced that some people used bad magic (*tsvera*) to steal from other people's fields. They were convinced of the existence of the magic because when they would herd cattle during the dry season they often found clay pots filled with water and medicines buried in the middle of particular fields. Why would anyone bury a clay pot filled with *muti* in their field, if not for the purpose of *tsvera*?[48]

---

[48] Incidentally some people thought that some people dug medicines into their fields to protect them from *tsvera*. As a result we can never be sure that the clay pots in people's fields were indeed *tsvera* or were an attempt to counter *tsvera*.

The father was convinced that *tsvera* existed because he had at one time been a victim of it. Someone who pretended to be his friend once came to his field during the evening to ask for snuff. When he left this farmer's field, this 'friend' collected a tin full of soil from his field. On discovering that this was what had been done, he followed the man to his field and found him still with the soil in the tin. He asked him why he had taken some soil from his field to which he answered that he wanted to use the soil to wash his hands in the river, which, as was later pointed out, he had already passed. The respondent was enraged and strongly suspected *tsvera*. He demanded that the soil in the tin be thrown away. In the end, the field never gave him anything so he had to change fields. The *tsvera* had already worked. It had made his soil cold. The official reason for moving from that plot was that the field was less than twelve acres. If he had asked the resettlement officer to be moved because the field had some bad magic on it, his request would have been refused. He believed that if there was enough rain, and a farmer had enough fertiliser, failure to get good yields would be a clear indication of *tsvera*. A prophet could confirm the suspicions if they were correct. The only way to neutralise *tsvera* was to use holy water.

Case 2

This farmer was a very poor farmer and people thought that he was lazy. He believed in the existence of *tsvera*. He believed that if you cultivated a large piece of land and applied enough fertiliser then you could expect to get a large yield. If the yield was low this would indicate the workings of *tsvera*. For him people with *tsvera* were like witches that operated at night and were able to get in even if you locked your door.

According to the respondent, some people also have magic that can make others unknowingly work for them at night when they should be sleeping.[49] The following day, the people who will have been overworked at night will feel too tired and lazy to work in their own field. Cattle can also suffer the same fate so that when their owners want to use them the following day they will be tired and sleepy. Faced with this problem, one can appeal to traditional healers and prophets. They will tell you who your enemies are and how to overcome their magic. When asked if his cattle had ever been used this way, the respondent said no, but when asked if he thought he had ever been a victim, he said yes he had been a victim:

> This year Farmers World gave us a loan for cotton. We had enough things for one hectare and we were supposed to get 7-8 bales per hectare but we got only 4½ bales.

---

[49] See Niehaus *et al*. (2001: 5) for similar accusations in a village in South Africa, where a villager was accused of keeping baboons and turning children into zombies to work in his garden at night.

To protect fields against *tsvera* people had to use protective medicines, which often turned around and started to 'eat' their own children. As a result, this farmer did not do anything to protect his field, though his wives sometimes got holy water from the prophets. He claimed that as a result of his possession by baboon spirits, he was not allowed to visit traditional healers (note that one of his wives was a traditional healer) or go to prophets. He strongly believed that in the village there were some people with *tsvera*. These people cultivated small areas but got huge yields. What was surprising to him was that all farmers loaned seed and fertilisers at the same time. He felt that, sometimes, he even worked harder than those whom he suspected of having *tsvera*, but always the culprit's yield surpassed the wildest imagination. His wife, a traditional healer, had once been accused of practising witchcraft on infants and toddlers at a village meeting. He refuted this allegation and maintained that it was because people hated him.

Case 3

The respondent was a very poor farmer and did not believe in the existence of *tsvera*. For him *tsvera* was a thing of the past when people did not use fertilisers and used cow manure instead. In those days a person would plant a small area and get a lot of maize, even more maize than those who planted a large area. This was because the person had *tsvera*. But the advent of fertilisers spelt the end of *tsvera*. For him, *tsvera* accusations are just like witchcraft accusations: when a person dies there is always a witch (*panofa munhu hapashaikwe muroyi*).

Case 4

This was a very capable farmer (judging from both local farmers' and AREX officers' standards). Neither he nor his wife believed in *tsvera*. The use of sufficient fertiliser was the answer to poor yields. The ability to plan and manage farming properly was the only *tsvera* that a person needed to be a successful farmer. However, their children, a son and a daughter, were convinced that *tsvera* existed, but they would only say so in the absence of their parents and they wondered why their parents were denying the existence of something they (the parents) knew existed. The children went further to comment on scenarios and examples within the community which to them indicated the use of *tsvera*.

As I consider these case studies I can't help but think of Zeitlyn's (1991: 61) comment that 'the existence of mythical creatures is more newsworthy than their non-existence', just like the existence of *tsvera* tickles my imagination more than its said non-existence. This can also simply reflect the fact that a lot of people in the sample believed in the existence of magic and that those who did not, or who at least doubted its ability to work, did not have much to say on the topic.

Magic is a complex issue since some people who denied its existence or ability to work often still took measures to protect their fields from people with bad magic. Even when the wife was the one looking for magic to protect the field, it was impossible to tell whether the husband was simply humouring his wife by allowing her to use magic; or whether he somehow believed in magic but was loath to admit this. Usually men left the duty to their wives, who took holy water from prophets and priests. All female-headed households except one had taken measures to protect their crops. Below are the views of one man and two women on magic.

> *Why do you not take measures to protect your crops when you believe in* tsvera*?*
> I am not interested. My wives are the ones who go to get the holy water.

On the issue of crop protection, one woman pointed out that:

> Some people put pegs (*hoko*)[50] in their fields to guard their crops against mysterious disappearance, but I use holy water from my church. When I go to Mutemwa this year, I will take a five-litre bottle of water. I want Sekuru Chakaipa[51] to bless it.

The second woman said,

> I use water that has been made holy by prayers from prophets in my church. I sprinkle the water in my field before I start planting crops so that my field is blessed by the power of God. If my field is blessed, so are the crops that come out of it. This blessed water will help my crops grow healthy and strong and will chase away any evil spirits so that when selling my crops I won't have bad luck.

There are gender differences in the way people consider magic. Men are more likely to acknowledge the existence of magic and then deny its effectiveness. On the other hand, in their role as providers of food, women need to feel they are in control over what happens to their food supply to ensure food security. They need good healthy crops to feed their families or else they risk having to go to *maricho* to work for food when there is a crop failure (men hardly ever go to *maricho*). The different domestic roles played

---

[50] *Hoko* is usually a peg, which is used to demarcate boundaries. It is usually a wooden stick or an iron rod but can be anything that marks boundaries. However, when used in association with magic and witchcraft, it refers to the herbs that are used but that are usually dug into the soil in clay pots or bottles.

[51] She belonged to the Roman Catholic Church and referred to Father Patrick Chakaipa (now deceased) as Grandfather (*sekuru*) Chakaipa because of his advanced age.

by men and women have an influence on their views on magic and ultimately on their behaviour. A big healthy crop is not necessarily associated with the variety of seed, although people realise that if you apply enough fertiliser and everything else remains constant, you are assured of a big harvest. Whilst farmers and scientists believe in the usefulness of fertiliser application, for farmers technology is good but needs to be propped up with a little magic to get good yields.

Often the good farmers are accused of using magic. All poor farmers except one believed in the existence of magic: all the better-off farmers ridiculed this idea. The farmers who were accused of using magic were sometimes accused because they were able to achieve extraordinary feats, such as cultivating only a small area but getting much more than people who cultivated twice the area and who worked twice as hard. Presented with such an argument, the wife of a farmer who was regarded as a good farmer and not considered by others villagers as using magic was adamant that,

> The only *muti* is the one I have been telling you about. If you plant the whole 12 acres with two bags of fertiliser that is, one top dressing, one down dressing, what do you think you will get? Then you start pointing to your 'next door' neighbour accusing him or her of using magic. You will be saying ah, why is it that I planted 12 acres and they only planted two but they got more maize than I did. It is because they had enough fertilisers for those two acres.

However, even among those who believed in magic, there was an acceptance that access to fertilisers could determine who the good farmer was. Regardless of this, most farmers who believed in the use of magic refused to consider the fact that they might have been incorrectly applying fertilisers to their crops. One farmer, who was convinced that he was always losing his crops to *tsvera*, stated that he had correctly and consistently applied fertiliser to his field. On a separate occasion, when he was asked how much fertiliser he applied to his crops, he maintained that he applied one bag of ammonium nitrate, and one bag of Compound D to every acre of land. This was despite the fact that, according to him, AREX recommended two bags of Compound D and three bags of ammonium nitrate per acre.

On the other hand, one of the good farmers who said that he used six bags of compound D and eight bags of ammonium nitrate per hectare, which was much higher than the village average for fertiliser application, pointed out that:

> In most cases if you work hard and have access to inputs then you are likely to get something. Others work hard, but because they do not have inputs and other farming implements, they do not have anything. You can plough your field and even manage to weed the whole field but if you do not have fertiliser you will not get anything.

However, the kind of farming skills a person has, or the kind of farming practices used and even the kind of seeds and technology applied, are usually not regarded as the first consideration in explaining success or failure. One female household head in Muringamombe alleged that in 2004 she had lost her maize to people with bad magic because by the way her crops looked in the field she had expected a good harvest but failed to get even one ton. This female head was convinced that her crops had been stolen using magic, despite the fact that she had planted an Open-Pollinated Variety on a large portion of her land and had not used enough fertiliser. There is an all-too-familiar pattern in which people look for external causes before evaluating their own culpability. People would rather blame someone else than admit that they do not have enough resources or that they are not applying the correct amounts of fertiliser or even cultivating low-yield, non-commercial varieties, whilst their colleagues are doing the opposite. As a result, the belief in *tsvera* can make people reluctant to question their own behaviour and sometimes they are even reluctant to learn new things from others since these others might be not be good but might be using bad magic. With all this in mind, one might ask whether *tsvera* is a defence mechanism for powerless and resource-poor farmers. But does not reduce the concept of *tsvera* to another idiom denoting powerlessness?

In discussions *tsvera* was mentioned in connection with food crops such as maize and sorghum but never in connection with commercial crops such as tobacco and cotton, which were regarded as modern crops. If accusations of the use of bad magic are reactions to modernity's malcontents or even a result of jealousy, why did these accusations not extend to tobacco and cotton, which could easily lead to jealousy since these crops were associated with more money and wealth?

Jackson (1989: 103) points out that some beliefs are held to be true because one's elders hold them to be true and hence they have the authority of custom. Jackson looks at a variety of factors that influence people's beliefs. For example, people can pick up beliefs from others around them. What others take for granted and believe, what people learn from parents and elders, what they learn from peers, hearsay stories, real experiences, etc. can influence a person's beliefs and ideas. This, however, was not always the case where *tsvera* was concerned. In the sample there were two families where the parents did not believe in the existence of *tsvera* but some of the children in the family did allude to its existence. Even for those who agreed with their parents that *tsvera* existed, the children often pointed to 'evidence' that was independent of their parents; as in case 1, where the children pointed to clay pots filled with water and *muti* in the culprit's fields. However, this does not completely rule out a belief caused by authority since what could be interpreted as evidence could well have a strong cultural underpinning.

The belief in magic worked against learning by observation. It was the general sentiment that during the rainy season people did not like people to come to their field because some would pretend to come to observe whilst they were putting in place their *tsvera*. This was highlighted in Case 1, in which the respondent maintained that his 'friend' pretended to come to the field to borrow some snuff when all he intended was to steal some soil so as to use magic to steal the respondent's crop. One informant in the sample was threatened with a beating in a nearby village for walking on the edges of someone's field during the rainy season, based on the suspicion that he was using magic because crops can follow the footsteps of the person with magic. Indeed on many occasions most people mused why anyone would want to pass through their field during the rainy season and, sometimes, staring at someone's field for a long time would raise the ire of the owner of the field. However, *tsvera* was not only used to steal crops whilst they were still in the field, but could also be used to steal grain from people's homes.

Familiars[52] that looked like dolls could be sent to *eat* the grain or a whirlwind could be sent to snatch the grain away. To deal with familiars one had to consult the prophet, while the whirlwind was easier. For example, when one was winnowing *rapoko* a whirlwind could be sent to steal it. According to one old woman, to make the medicine powerless people could lift their index finger and say something like 'take the *rapoko* and see if you will benefit'. The whirlwind would 'feel shy' because it would know that it had been seen and so it would not take the *rapoko*.

All people who believed in *tsvera* magic agreed that there were ways to counter the magic. However, counter-magic was always risky as it could turn from being good to being evil. As a result, most people claimed that they did not protect their fields with counter-magic but sometimes used holy water blessed by the priest at church. Even those who used *tsvera* were at risk since at any time the *tsvera* magic could turn against them. With this in mind, not all the good farmers were accused of using magic. Only those with poor households, that is those who got bumper harvests but had nothing of material value to show for it, were accused. Added to this category were good farmers with problematic homesteads. For example, a mentally retarded child was taken as a sign of the use of magic by a farmer whose child was afflicted in the same way. Is it legitimate then to question whether

[52] Witchcraft familiars refer to the tools that the witch uses to perform his/her witchcraft acts. These familiars can be inanimate objects such as dolls or herbs to which people attribute the human-like abilities to move, think and act. Familiars can also be dead people who are taken from their graves by magical means and controlled to do bad things by their owners. In some cases familiars are understood to be things made by witches, which acquire human form although they always take the form of dwarves or young children. Snakes, owls and hyenas can also be witchcraft familiars.

the fact that people do not have total control over farming inputs and implements explains their belief in *tsvera*?

This belief in magic also impacted on people's willingness to discuss issues related to their yields with other people. If their crops failed to fetch a good price on the market, and they had little money to spend after a good harvest, they were likely to be suspected of using witchcraft. This is so because it is usually believed that money obtained by using magic is like paper and cannot buy anything worthwhile. In some cases the goblins which are used to amass such wealth are very demanding such that the money is then used to purchase things for the goblins to keep them happy (when goblins are not kept happy they will start causing havoc for their owner).

Such beliefs restrict the spread of information about seed varieties. It also impacts on the free circulation of seed within the same village. People exhibited great mistrust of their fellow villagers and instead felt comfortable carrying out transactions with people from other villages. For example, it was believed that some people could give you seeds that they had treated with *tsvera* medicine so that if you planted their seed then your yield would automatically go to their field. Alternatively, a person could give you seed which could turn into witchcraft familiars and haunt you and your family. This sentiment might also be a factor in building social networks, since people always pointed out that their best friends, with whom they exchanged information and other things, did not live in their village.

Magic is not always used to steal from others or to protect crops against people with bad magic: it is also used to protect crops against marauding animals. To do so, people have to observe certain associated taboos. For example, according to one traditional healer, some people have rules not to take salt to the field, so as not to neutralise the magic that protects the field. They should not eat any of the field crops whilst standing or walking. Either they have to sit down and eat at the field or carry everything home and eat on arrival. However, before they are allowed to pluck anything from the field, monkeys should be given their share, after which no animal will attack the field. One woman would hang some cobs of green maize in trees around the field for the monkeys to eat. This was to thank them for not eating the maize. This was done because the woman had magic to protect her field against marauding animals but for the magic to work it was stipulated that this action had to be carried out every time before eating any of the first crops from the field. As a result, farming knowledge not only meant good land husbanding methods, but also good magic that could make the difference between being a farmer and being a very good farmer.

Witchcraft

It is difficult to offer an accurate definition of what constitutes witchcraft. In Shona the word for witch is *muroyi*. A *muroyi* is basically an evil person who specialises in killing people through magical and mystical means, which are associated with the use of bad *muti*. Gelfand (1956) also regards killing a person by poisoning as an act of witchcraft. According to Bourdillon (1987: 183),

> ... witchcraft is the paradigm of all evil and anti-social behaviour, but not all such behaviour is witchcraft ... witchcraft can refer to any threat, involving an element of furtiveness, to personal security by the violation of the human person or of human life, or by the violation of any deeply held value. A witch or sorcerer (*muroyi*) is a person in any way responsible for such a violation.

Some informants referred to witchcraft as a 'science' and sometimes white man's science was referred to as witchcraft. However, witches always intended harm on others whilst some good could come out of the work of the scientist. The similarity of the two lay in the fact that the workings of scientists and witches were mysterious to ordinary people. That is, the 'witchcraft of science', and the 'science of witchcraft' are not easily understood by lay people.

On the other hand, witchcraft is not so very different from magic – it can be understood in some cases as a form of bad magic – with the *proviso* that magic is not always bad and witchcraft is invariably directed at harming others. Hence, when magic is used to harm others it is sometimes referred to as witchcraft. For example, some people in the sample referred to *tsvera*[53] as witchcraft. The belief in witchcraft sometimes took weird twists. Getting a very good yield when everyone else did not get anything could sometimes be a shaming and embarrassing experience. One farmer, who achieved an excellent groundnut harvest, whereas everyone else in the village got nothing, was embarrassed to shell her groundnuts during daylight and took to doing so at night. She could not even ask people to help her because she was afraid that people would accuse her of stealing their crops by using magic. She feared people could easily suspect her of witchcraft since she had previously lived with a *n'anga* (traditional healer). However, a large crop does not always result in witchcraft accusations. Accusations depend on a number of factors. If it is generally a good year for everyone then people are less likely to be accused of witchcraft, though in a bad year, or when other

---

[53] Of Gelfand's nine different kinds of witchcraft, I am mostly interested in his definition of *divisi* which is also another name for *tsvera*. Gelfand (1956: 53) defines *divisi* thus: 'There is a witch who possesses a *gona* or special horn filled with medicine which has the power of producing plentiful crops. The power of *divisi* passes to whoever possesses the *gona*, but before the bountiful crops can be obtained the *gona* must cause a death or even several deaths … The fact that he (the owner) accepts the price of procuring such a *gona* makes him a witch or tantamount to a witch'.

associative factors are considered, like this woman's previous association with a traditional healer, good fortune might be viewed with suspicion.

My respondents differentiated between two kinds of witchcraft,[54] *uroyi hwedzinza* (inherited witchcraft) and *uroyi hwekutenga* (when a person is not inherently evil but buys the evil charms and portions, thus corresponding to Bourdillon's 'sorcerer'). But there is a fine line between the two, since the person who buys the witchcraft paraphernalia can pass it on to his or her offspring, resulting in inherited witchcraft after all. For some respondents *tsvera* was also witchcraft and was associated with *uroyi hwekutenga*. A person who is not inherently a witch can buy the *tsvera* magic from corrupt medicine men or women, hence the popular belief that the *n'angas* are also witches.

For the medicine to work, the person who has bought the *tsvera* might be asked to sleep with his daughter or to marry her daughter to a goblin (*chikwambo/tokoloshe*). If this is done the daughter will never marry and sometimes cannot enjoy relations with men. The father of such a child is branded a witch since his greed for riches is seen to cause his daughter to suffer. Sometimes sons can also be afflicted in this way if their father owns a female *chikwambo*. In some cases, as discussed with the respondent above, the children of the person with *tsvera* become zombies. Some people believed that one of the good farmers in the village used witchcraft for farming because he had a mentally disturbed child. Anyone accused of witchcraft is accused of practising evil, maybe to their own advantage but certainly to the detriment of everyone else. That is why magic such as *tsvera* is regarded as witchcraft: it can harm others. In witchcraft discourse the idiom of exploitation is rife so that witches are accused of using some people to achieve certain ends. For example, people often referred to the unmarried daughter or the idiot child as such an *arikushandiswa* (someone who is being used). One woman said they would welcome a witch-hunter to flush out the witches, some of whom could be identified

> ... by the fact that their granaries usually burst open at certain times of the year and when they burst you certainly know a member of the family is going to die. This is so because the *tokoroshis*[55] that are used for farming demand payment in

[54] Bourdillon (1987) contrasts between witches and sorcerers. For Bourdillon (1987) witches kill people and do other evil deeds because they take pleasure in seeing people suffer. On the other hand, sorcerers buy their evil charms from evil herbalists and then use these charms to exact revenge on others, or for personal gain to the detriment of others.

[55] *Tokoroshis* are another kind of witchcraft familiars. They usually take the form of dwarf-sized individuals. These *tokoroshis* are rumoured to be imported mainly from South Africa, hence the word *tokoroshi* is a derivative of the South African word *Tokoloshe* that indicates the same. *Tokoloshes* are very powerful and ruthless and love human blood.

> the form of human blood. And they should get this blood every year. These deaths usually occur at the same time every year and the sequence of events is the same first, the bursting of the granary and then the death..

When I attended the village court in one of the villages, I noticed that all the people seemed to defer to one man and he seemed to command a lot of respect. Later I asked some women why people respected him so much and they told me that it was not respect but fear. People thought he had bad *muti*. This man was believed to have witchcraft powers to eliminate anyone who displeased him. People also thought that he used *muti* for farming. He cultivated a lot of cotton and maize and even employed people to harvest for him, but he did not have property commensurate with his farming prowess. However, I had an occasion to talk with the feared man who complained bitterly about the pricing of his cotton crop by CMB. 'I sent twelve bales and they were all graded C. Getting $3,000 ($55) for a bale! What am I farming for?'

Was he really a witch or was he just a victim of unfair pricing policies? In one instance, cases of infant illness and deaths, which had been occurring in one village, were attributed to witchcraft. At a village meeting during which all this was discussed, men accused women of witchcraft and most of the people pointed indirectly to one female traditional healer. They all agreed that the person who treated children for a fee was the one who was bewitching the children, so that they would go to her for treatment. (The one-year-old son of the people I was staying with was also attacked by *chivere*. Fortunately he lived after the mother gave the child some traditional medicine and performed a ritual that her mother had taught her was effective against *chivere*.)

Others said that prophets at the church had advised them not to take children below the age of two to the borehole because that is where evil medicine was located. The prophet had said that if nothing was done to remove this medicine a lot of children were going to die from the sickness which people referred to as *chivere*. At the meeting, other child 'doctors' stood up to defend themselves, saying they were clean because they did not charge money for treating the children. People chanted slogans about how children should be loved because they were innocent souls.

Of the three women who treated sick children, two defended themselves. As it turned out later, the husband of the third woman who was thought to be the witch did not sleep that night, and accused all the people in the village of having accused his wife and at a subsequent meeting threatened to sue people in the village.

At the meeting people proposed solutions, none of which were taken seriously. Some wanted to invite Tsikamutanda the witch-hunter, some wanted to bring a prophet to remove the bad medicine. Those who opposed to inviting Tsikamutanda were darkly accused of knowing something. In later conversations with other people it also emerged that they believed the third healer was a witch because she had proceeded to a *dandanda* (a traditional ceremony) even though she had been told that a child of their *sahwira* had died of the mysterious illness (*chivere*).[56] Leaving the dead child of their *sahwira*, not attending the funeral and going to a *dandanda* instead was unforgivable. According to the people, it was behaviour typical of a witch. Even after the incident, the said witch and her *sahwira* remained very good friends. As it happened, after his cattle had been threatened by black leg, the man whose child was said to have been bewitched by the healer promised to give his *sahwira* medicine to vaccinate the cattle as soon as he received some from his sister.

Some people in the village had very good relations with the witch and her family and discussed many things with them, including agriculture. However, on several occasions I was warned not to become too friendly with the said witch as she could take advantage of the friendship and bewitch me. If people did not discuss things with others, it was usually not because of witchcraft but because the people were seen as snobbish. Thus, although witchcraft could limit the flow of information, it was not always a major consideration.

Witchcraft accusations sometimes take place when people are already involved in conflict or when people want answers to occurrences that do not immediately make sense to them. For example, one woman used the idiom of witchcraft to explain why her family was in economic decline and why they had stopped cultivating some crops they had cultivated prior to coming to the resettlement area. She said that she had stopped cultivating crops like sorghum since she had been ill for a long time. Below is a conversation that took place between Christine and the woman:

> When I came here I worked very hard then one day I had a dream. I dreamed that I had been hit by a magic horn. A voice then said to me: 'I have hit you with this magic evil horn because you have gone to the resettlement and are going to be more successful than us'.
>
> *Whose voice was it?*
> I saw the person. He was my husband's relative. I was given his whole face in the dream and I knew the voice.

---

[56] I have to state here that I was very relieved when I was not pointed out as the witch. I was still new in the area and the illness more or less coincided with my arrival. I was afraid I would be accused and this would put my research in jeopardy. However, as I learned later, this illness was not new to the village, so the witch could not have been a new person. I was also happy that people did not call Tsikamutanda because then everyone would have been forced to drink his witch-finding medicine called *muteyo*. The media reported that sometimes people died after drinking this medicine.

(I remained quiet for a moment to regain my composure. I wanted to laugh. I could not believe the story.)
Yes I was given the face. When I woke up in the morning, I was very ill. Around this time, I vomited the horn.

You vomited the horn. Had you been treated or what?
No, I had not been treated. It is something that just happened. I just felt like vomiting so I went to those trees.

Ho-o on, that guava tree (the tree behind the main house).
Yes in between those two trees. Between the guava and the orange tree. I just felt like vomiting and I vomited the magic horn.

Ah! (In surprise)
The horn was like this size. *(Indicating the size of the horn on her fingers. The horn was indicated as being half the size of her longest finger*). The horn had a chain of beads around the neck ... The beads were of the white and black colour ... The whole village came to see the magic horn with its beads ... I have been suffering for a long time now. Since I came here, I have never been able to do anything.

When they first came to the resettlement area this family experienced a marked improvement in their lifestyle. They were able to purchase cattle and other farming implements. However, recently they have been faced with economic decline. The household head has become very old and partially blind and can no longer work in the field. Two of his daughters died after suffering from AIDS for a long time, which disturbed the farming activities of the household. At the same time, the twelve acres which the family started out with have been reduced due to subdivision of the field as the family grows. As in most other families, this fragmentation of farms became necessary due to the various conflicts which arose where parents shared the same field with their adult children. In some cases conflict was latent, whilst in others conflict became openly violent to such an extent that parents were even beaten up or accused of practising witchcraft on their children. As a result of the increased fragmentation, people have had to make trade-offs between the various crops they grow. People with highly fragmented fields also resort more and more to growing maize to meet the food needs of their families, unlike in the past when they used to grow more cotton. The husband of the woman above had this to say:

> I gave my children 2 acres each to grow their own food. We decided to subdivide this plot in the 1999 to 2000 season because we were having problems with combining our labour. From the 1999-2000 season I stopped growing cotton. I am now concentrating on maize because it is a food crop. However, on his two acres my youngest son grew cotton only.

This family could not focus on cash crops like cotton, which used to give them cash to purchase property. Maize was not regarded as a cash crop but solely as a food crop. At the same time, most of the food they produced was needed to feed the young orphans they were left to take care of.

We can also understand how farmers make decisions to either or not adopt new knowledge or technology by looking at associated indigenous knowledge, especially with respect to beliefs in witchcraft and medicine. For example, in cases of *chivere*, raised earlier on in the chapter, people would resort to traditional healers and prophets, who confirmed their suspicions that the illness was caused by witches. Villagers also agreed on who was causing this terrible illness. Even village health workers were convinced that it was *chivere*. What is striking is that no one suggested calling in the health personnel to determine the scientific cause of this illness, which they said always attacked children in the village at certain times of the year. No one was interested in considering alternative explanations or verifying whether what the healers and the prophets said was true, because the prophets and healers had confirmed what the villagers knew to be true.

Parallels can be drawn with the way people adopt or refuse to adopt new technology. For example, some people were disgruntled by the blue fertiliser which they had received from Farmers' World in 1999. They said the blue colour of the fertiliser was washed away by rains and left behind white stones. People were also particularly disappointed with the *katsoko* seed that they had received from Farmers' World the same year. When their crop failed, people blamed *katsoko* and the blue fertiliser. Upon further questioning it emerged that there was more than average rainfall that season, but no one ever considered the fact that the fertiliser might have been leached by the rain, and only one person linked the failure of *katsoko* to the excessive rains. AREX' explanation was that they had told people that *katsoko* was an early maturing variety and that if people failed to get their seeds early they would have to plant the *katsoko* in early December. Whether it was for lack of understanding or love of experimentation, people planted *katsoko* very early at the start of the season. When the seed failed they turned around and blamed AREX. Just like in the case of the mysterious illness, people did not further investigate the causes of their misfortune. Therefore, the adoption of new technology is not simply based on its technical qualities. Rather it is based on the schema which individuals and groups use to verify their knowledge. In this case, farmers used indigenous systems of thought to verify western scientific technology. Verification and production of knowledge do not end with the scientist. If the introduced technology does not fit into the existing patterns of knowledge and beliefs, people might resist it.

The effects of witchcraft rather than a physical proof of its existence are seminal in accusing people of using witchcraft. People deny that they learn things by observing the fields of neighbours. They say that if you show too much interest in the fields of those around you, people will suspect you of witchcraft. While this is a widely accepted observation, as discussed in the chapter on experimentation and observation, some of the knowledge that people have is in fact gained through surreptitious observation of neighbours' fields. However, the impact that the belief in witchcraft has on knowledge is such that the circulation of knowledge within the village is inhibited since people are usually freer to discuss matters with those who live in other villages than their own.

Some of the informants believed that sea-shells[57] (*nyengeresi*) were very dangerous, as they could be used to bewitch a person. One young woman who had three young children maintained that *nyengeresi* was used by old women to cause illness in young children. Even adults could be bewitched using *nyengeresi*. Old women were especially suspected of using seashells. The person who would be bewitched by such potent witchcraft would start wasting away as if s/he had AIDS. At funerals these old women with *nyengeresi* would hold other people's children with the intention of making them ill, so that the parents of the child who became sick because of contamination by *nyengeresi* would have to go to them to pay for the treatment of their sick child. The old woman would pretend to treat the child while all she would be doing would be washing the child with *nyengeresi*. Although this story was related to illness and health, the message instructed people to distrust everyone - the AREX officers and their scientists and even their own relatives. The only person one should trust is oneself. Beliefs in witchcraft lead people to carry out their own experiments and come up with new knowledge that is relevant to their needs; however, it prevents them from learning from the experiments of others. This lack of trust ensures that experimentation does not end at the gates of agro-industry. Hence people are not simply passive recipients of knowledge but are also active participants in its creation.

Religion

In this section I discuss Shona religion and Christianity. With regard to Shona religion I also discuss the issue of *chisi* (rest days). Below is a summary of what both religions entailed for my informants.

---

57 In a similar study in Wedza area, Bourdillon *et al*. (2002) recorded that one man claimed to protect his field from *tsvera* magic with *nyengeresi*, while some villagers said that this man was a witch because he used the sea shells which were believed to be detrimental to people's health, and people believed that this man used magic to steal their crops.

Shona religion

Bourdillon (1987: 237) correctly points out that among the Shona, 'religion is concerned with persons who lived in the past and their supposed control of present events, religion serves to bring the past and the present together promoting in the living community a respect for tradition'. This assertion is valid at both the family and the community level. The Shona believe that the spirits of the dead come back to look after the living and to take care of their health and wealth. To demonstrate the link between the dead and the living on the question of wealth and general well-being, one informant responded as follows:

> *What causes people to be wealthy?*
> Some people say that their ancestors are the ones that give them wealth. It is because they appease their spirits. Some people get rich in their old age. Like Chidavaenzi. He was a very rich man before he converted to Christianity. He left his traditions and joined the apostles. He threw away what his ancestors had given him. When they converted him, they gave him six rank badges to denote that he was now the leader. Now he has nothing. He was the first person to own buses here. They pretended to like him when all they wanted was to drink tea at his home as he was a very rich man.
>
> *So we can say the ancestors can make people rich?*
> Sure. When we migrated from Bushu to Goora, we were very poor. We used other people's cattle for farming, then my brother worked hard and the ancestors blessed him. If you go to keep 7 and ask for Abraham Pfunde, everyone knows him. He built a shop with a bar and butchery. If we go to his place, we are not afraid to eat anything he offers, which is plenty. We say to ourselves let's eat. We were once poor. Now he wants to buy a house in Bindura so that he can sit all year round like this. (He crossed his legs and cupped his chin in his hand.)

The spirits have to be honoured and appeased to be happy. Otherwise things will not go well for the person who displeases them. Even when it comes to farming, if the ancestors are not pleased with your conduct, they can make sure your crop will not do well or they can give you bad health until you do their bidding. Thus, a poor harvest was not always attributed to poor application of technology, or even to losses due to bad magic, but could also be attributed to ancestral spirits. Some families mentioned that before planting their seed they would leave a portion of their seed to spend the night at the *huva* (an earth bench in the kitchen where claypots[58] and water buckets are kept) for the ancestors to bless the seed. When people pray to their ancestors they usually do so in the kitchen facing the *huva*. In the case of a

[58] For the symbolic association between clay pots and fertility see Herbert Aschwanden (1989). Although his insights are too Freudian for my liking, his analysis gives insight into some of the symbolism that affects fertility rituals, not only among the Karanga but among other Shona groupings as well.

death in the family, the body lies in state at the *huva*. Aschwanden (1989: 99) refers to the *huva* (or *chikuva*) as an altar or sleeping place of the ancestors. The relationship between the ancestors and the fertility of the fields is represented by the act of leaving the seed at the *huva* overnight for the ancestors to guarantee fertility.

In one village, one of the brothers of the village head had passed away the previous year. This deceased brother was of the Marange Apostolic church and did not believe in participating in the traditional rituals that the *Sabhuku* (village head) upheld. Traditionally, when an adult dies, at his funeral people have to slaughter a cow/ox which they call *nhevedzo*. In this instance though, the *nhevedzo* was not slaughtered at the funeral. This was rectified a year later because of some problems which were blamed on this omission. Many people we talked to in the village about other issues always found a way to include the case into the conversation by alluding to *mupostori wekufa gore rakapera wekuzobairwa mombe gore rino* (That apostle who died last year and had his *nhevedzo* slaughtered this year). This case was used to demonstrate that spirits are very powerful and that, even if you are a Christian, you still have to do their bidding. For some reason people found the story quite funny because they always joked about it during discussions on religion. Thus, all spirits of the dead in the family have to be appeased according to custom. Any omission with respect to this can bring disaster to the family, should the spirit of the deceased be angered. The disasters can range from bad luck in business dealings to poor health or a poor crop yield.

At the community level, the dominant figure in the religion is the *mhondoro* (lion spirit[59]). In the study area, the *mhondoro* spirit that was frequently mentioned and regarded as powerful was Nyamaropa. The land was regarded as belonging to Nyamaropa. Some apostles were adamant that the land belonged to no one but God, but they still took part in activities to honour the ancestors, though not always very actively. When asked why she could not just decide to flout the culturally prescribed *chisi*, a woman of the apostolic faith linked the land to the lion spirits. She said it was out of fear that if she went to the fields on *chisi* and something happened people would blame her and ask her if the land belonged to the apostles. Furthermore, she maintained she would not loose much by not going to her field one day a week.

These lion spirits are the owners of the land and are responsible for securing soil fertility as well as for ensuring rain. There are two types of *mhondoros*. There is the lion spirit of the earth (*mhondoro yepasi*, who controls soil fertility) and then there is *mhondoro yemvura* (the lion spirit of water). The lion spirit of water (rain) is said to be the lion spirit that went

---

59 For a detailed description of Shona lion spirits see Bourdillon (1987: 253-82), and also Gelfand (1956: 11-34).

into the water while it held some soil in its hands. Although it was immersed in water neither the lion spirit nor the soil in its hands got wet. This is the spirit that would later return to live among us as the lion spirit of the water. On the other hand, the Lion spirit of the earth is usually quiet and does not say much. One old man said that the difference between the lion spirit of the earth and the lion spirit of rain was that the lion spirit of rain (water) demanded rain-making ceremonies every year. However, the lion spirits of the earth only got *chisi* days when they could walk undisturbed.[60]

The lion spirit is usually the spirit of an elder who died long ago. The spirit dwells in a wild lion and can possess a medium who is known as *svikiro* (the place of arrival) or *homwe* (pocket). When the lion spirit comes back to dwell among people it should possess someone who does not come from the village but from far-away lands. This is so because the person who is possessed should be able to tell the villagers new things about the village, things that the villagers might not even know, but that are nevertheless true. A true lion spirit does not arise from the same village but comes from another place.

Also, like family spirits, the lion spirits of the land may also be responsible for the wealth of the people in the community, providing these spirits are honoured properly. 'If the descendants of the *mhondoro* obey his laws and perform ceremonies in due time, they will live in peace and plenty' (Lan 1985: 32). Chief Bushu's wife noted that even the white commercial farm owners respected the spirits and informed the spirits if they wanted to build any structures on their farms, otherwise the built house would become uninhabitable or haunted, or would simply fall down. She pointed out that even the gold miners bring gifts of snuff and cloth for the spirits so that they can obtain the gold, otherwise they will get nothing. To this I replied that my Father once had a gold claim at Chin Mine, and that he did not get any gold and went broke because of it.

> Maybe he had not done the proper rituals for the *mhondoro* spirits such as giving the cloth and snuff.
>
> *When I heard later about the* mhondoro *I asked him whether he had done any such thing and he told me he had.*
> Maybe he did not do it correctly, or there were some bad people around him who did not want him to get the gold. It is also possible that he did not know where the gold belt was.

The lion spirits maintain their territorial integrity by guarding their land and their wealth jealously from outsiders. Any misfortune suffered by these outsiders, whether in terms of poor crop yield or, like in the case of my

---

60 Some people hold the opinion that the lion spirits of the water also walk around on *chisi* to check whether adequate rain has fallen.

father, a missed gold belt, can be explained as a failure to follow the proper ritual, and not by a lack of knowledge on how to deal with he new environment.

Since the lion spirit is the guardian of fertility and wealth of the land, all the people who dwell on that land and hope to reap from it should pay homage to them. Thus, regardless of race, wealth and religion, everyone is supposed to contribute to rain-making ceremonies or other religious ceremonies as the spirits might demand. For the rain-making ceremony all people are expected to contribute *rapoko* and maize. The apostles who are loath to take part in traditional ceremonies still make their contributions to the village head.

One woman who had previously lived and worked on Large Scale Commercial Farms run by whites maintained that if there was a drought the white men would go to the *mhondoro*. They usually went to Musana or Chikwakwa. All the white farmers would contribute different things, bags of maize, *rapoko*, snuff, *nhekwe*, wooden plates, and take these to the lion spirits to ask for water. The lion spirits would then ask the old women to brew the beer. On the appointed Saturday the farm workers would be released early to attend the rain-making ceremony where they would spend the night drinking beer and eating meat. This woman, however, was of the Marange apostolic faith, so she never actually attended these functions. Only old women past child-bearing age, who are no longer having sexual intercourse with their husbands or indeed with any other men, are allowed to brew the beer. Aschwanden (1989) has presented accounts of how semen is synonymous with dirt when certain traditional ceremonies and rituals are performed. Therefore, some spirits forbid their female hosts to have sexual intercourse so that they (the spirits) are not exposed to the dirt. Pre-pubescent girls are the only ones who are allowed to fetch the water used for the beer brewing. It could be that pre-pubescent girls are mostly virginal as they are not yet ruled by the hormones that tempt them into sexual relationships with men. Menstrual[61] blood is also regarded as unclean and any menstruating woman would spoil the beer as that would become unclean as well, and the ancestors would be displeased.

The beer is brewed in the forest at a hut called *zumba* and is drunk in the forest at the *zumba* and nowhere else. During the ceremony two pots of beer are left at the *zumba* for the lion spirit to drink. After two days the people go back to the *zumba* to finish the beer that the lion spirits will have left. The lion spirits do not drink everything but only drink some beer from each pot. The lion spirits have never been known not to drink the beer.

---

61 The fear of menstrual blood is dominant among the Shona. Menstruation is associated with a lot of taboos. For example, it is believed that if a menstruating woman walks through a groundnut field the groundnuts will not bear fruit, or if a menstruating woman shells maize the maize will be easily attacked by weevils.

The *zumba* is not only associated with rain-making ceremonies but also with fertility. People take a portion of seed from all the food crops they cultivate to the *zumba*. These seeds spend the night at the *zumba* where the lion spirit blesses them, and then people can mix these with the seed they eventually want to plant. However, according to the village head of Mudzinge, treated seed and modern crops like tobacco and cotton can never be taken to the *zumba*[62] because they are sprayed with smelly chemicals and the *mhondoro* will not recognise the crops.

In an attempt to establish the efficacy of the rain-making ceremonies, I had the following interview with the Sabhuku who was the leader of these ceremonies. However, the discussion brought to the fore some other issues related to the lion spirits. On being asked whether people thought the rain ceremony really worked, the Sabhuku pointed out that

> ... people believe it works. They are actually troubling me to start the process. Now there is Mashaga, a village head in Takawira. He announced that he is going to start his own *Zumba*. So I am afraid that if I brew the beer he will poison it since it is left alone in the bush. If people get poisoned they will blame me.
>
> *What are others thinking of this issue?*
> Most people were angry. They asked me to take him to court over that issue but I refused. What had given him that thought? When we first came, the lion spirits came here and showed us the place where they wanted the *Zumba* to be. If he wants to go against the spirits, let him, but I am not going to say anything. If the spirits want to punish him, they will do so. Now if I was to confront him and he fell ill people would begin to think I did something to him. But if the spirits just punish him on their own volition people will tell him that the misfortune is a result of his own folly.
>
> *So you are not going to brew the beer this year?*
> No we are not. Kaseke had suggested that we should go ahead and brew the beer but then make sure that people sleep at the *zumba* to guard it but I refused. One day there might be a lapse of security and the beer could be poisoned. Even on the day people drink, someone might be bribed to poison the beer and all the blame would be on me. No, we are not going to brew anything.

People from the other village understood the matter differently. As it happened, a man from one of the study villages had found a pangolin in the bush. As is the custom, the man did not eat the pangolin but decided to give it to the chief. He took the pangolin to the headman who led the-rain making ceremony for him to take the pangolin to the chief. The village head promised that he would take it to the chief but instead ate the pangolin

[62] Lan (1985: 47) found the same consideration in the Dande area of Zimbabwe where he carried out his research on the role of spirit mediums in the liberation war.

himself. When he got no thanks from the chief, as custom requires, the man who had found the pangolin went to the chief to inquire about it. When the chief knew that the village head had eaten his pangolin he demoted him.

The spirits can be political.[63] The issue of the *mhondoro* is highly political. When the village head lost the power to be the leader of the rain-making ceremony it was not only a loss for himself but a loss in status and prestige for the village as well. That could explain why there was a suggestion to go ahead and brew the beer instead. There is also evidence that tradition can be circumvented and changed to suit current needs. For example, I would have been interested to know how the lion spirits would drink the beer at the *Zumba* if the villagers had guarded it all the time like Kaseke suggested. The issue that the spirits have power over people's health and fortunes was also highlighted when the village head remained convinced that the spirits would punish the other headman who had taken over the rain-making ceremony without sanction from the lion spirits.

Even in traditional religion people have to guard themselves against charlatans and a 'good' or 'real' *mhondoro* is recognised by its power over nature and lesser spirits. For example, in a discussion with Mr. Virimayi it turned out that a person whom others thought was possessed by goblins claimed that he was the medium of Nyamaropa, the highest spirit of the land. Some people believed him and built a thatched little hut for him along the river. According to Mr Virimayi, it later turned out that the child was possessed by goblins when the real Nyamaropa came and chased the child away. He maintained the Nyamaropa punished the charlatan as he is now very poor and does not have anything to his name.

> You know Nyamaropa can do it. He is the real thing. That impostor was chased away. Even if you do something that is not permitted, Nyamaropa will bring you to heal. If you plough on Friday, you will get into trouble.

---

[63] According to Ranger (1967), among the Shona people the spirits played a very important part in mobilising people to fight against colonialism. Spirits like Nehanda and Kaguvi led the first attempts to fight against white colonialists in what is referred to in literature as the first Chimurenga of 1896. *Mhondoro* spirits have played an active role in Zimbabwean politics. Lan (1985) documents the role played by the *mhondoro* in organising and mobilising people during the second Chimurenga war which culminated in independence in 1980. In the current highly politically contested land invasions in Zimbabwe, some lion spirits have also been at the forefront of claiming white-owned land that falls within their domain. However, Beach (1979: 419) argues that the political role of the spirits, especially in the 1896-7 uprising, was overemphasised. Instead, he points to the spontaneity of the rebellion against the colonialists without the intervention of the religious leaders. Beach (*ibid.*) maintains that Ranger's thesis, attributing the rise of the 1896-7 revolt to a pre-planned revolt led by religious leaders, resulted from Ranger's misquoting of sources.

(Netsayi) Did any misfortune befall someone in this village who had disobeyed?
Yes (*he points to Josiah's homestead*). He ploughed on Friday and as soon as he finished, the cattle he had used were eaten by lions.

Why did he choose to disobey the rules?
I do not know and incidentally he is from the Nyamaropa family. He just refused to listen and said that because he was apostolic he could do anything he wanted.

So Nyamaropa is serious. If it had been someone else they would probably have spared their relative and given him some other form of warning.
As a descendent of Nyamaropa he should have known better. If the Nyamaropas disobey the rules, who will follow the rules? If you just follow the rules nothing bad will happen. I have not heard of anyone whose animals were eaten by wild animals or beaten by a snake.

I heard that one of Chakupadedza's grandchildren died from snakebite.
Ho that one. They know what happened. They know. A snake bit the child but it was not a real snake.

So what was it?
That snake was not God's snake. It was sent. The father of the child was demanding custody of the child since he had separated from the mother. The father sent for the child but Chakupadedza refused to let the child go. He told them that if they continued to refuse a snake would bite the child but they refused to believe this. The father of that child sent that snake so that neither of them would have the child if the child died. That snake was not a real snake.

The question arises why people so easily believed that this was Nyamaropa. It could well be that people in that village were happy that Nyamaropa had come, as this would raise the status of their village. Their village would have been the centre of focus for all traditional rituals. Indeed one of the people in my sample, whose village was next to Rataplan, was convinced that people had used magic to steal the lion spirit from the young man since they knew that being possessed by the lion spirit would give them power. The above excerpt also brings to the fore the fact that the lion spirits have domain over all living things. For example, wild animals would not attack one's animals or crops if one followed the laws of the land, which are invariably the laws of the *Mhondoro*. The *Mhondoro* cannot allow animals under its domain to harm the people under its protection, hence if a snake bites anyone then it cannot be a snake at all but is rather some magic sent by a person with evil intentions that takes the form of a snake.

The *mhondoro* is well known for disliking modern things. For example, according to Chief Bushu's wife, the lion spirit had to be persuaded to allow people to build houses roofed with zinc sheets or asbestos sheets. When she was asked to explain how change was possible if the ancestors who disliked modernity and change were also part of the decision-making process, the chief's wife pointed out that some changes are made after the death of a spirit medium before a new one takes its place. A clever Chief takes this opportunity to make a few changes because when the new medium comes he or she will usually not challenge the chief's decisions, as it is thought that the chief always makes inspired decisions. Sometimes the *mhondoro* can succumb to logical arguments:

> For example, the lion spirits here did not allow people to use zinc roofs or asbestos sheets because they said the material was shiny and the shine would flash into their eyes blinding them (*tinozopenyerwa*). I think it was in 1965 when we put asbestos sheets on our house because my husband went to the *mhondoro* and asked him why he did not want asbestos sheets and zinc roofs. My husband then pointed out that people were no longer using clay pots but shiny modern pots and tins to fetch water. Did it therefore mean that all those things had to be thrown away if the *mhondoro* did not like the shine? After a lot of debate and discussion, modern roofing was then allowed.

Chisi

Chisi is a sacred day devoted to the spirits, during which agricultural work in the fields is prohibited. That is, *chisi* is a mandatory day of rest. 'Failure to observe *chisi* would result in the failure of rains or crops destruction by animals' (Matowanyika 1991: 230). According to Bourdillon (1987: 70) *chisi* expresses the association between the spirits and the land. This is aptly shown in the quotation below where one respondent was asked why people could not go to the fields the day after the first rains: 'Because the lion spirits will be walking around inspecting fields to check whether the rain received is enough. If it is not enough we will get more rain again'.

Sadomba (1999a: 35) neatly summarises the relationship between land, spirits and people for the Shona: 'Nature is close to the spirit world and through it the spirits and God communicate to the humans'.

This relationship between the spirits and the land is very immediate. If the spirits are unhappy, the land does not produce. For example, the rinderpest disease that killed cattle and the outbreak of locusts that led to famine in southern Africa (southern Rhodesia included) in 1896, were interpreted by Africans in southern Rhodesia as a show of rage by ancestral spirits who were angry with the people for not fighting against the European invaders (Zvobgo 1996).[64] If anyone hurts the land in any way the spirits are unhappy, and the land does not produce and rivers and ponds can dry up. We cannot, however, talk only of an association between the spirits and the land since the spirits are the land and the land is the spirits. It is difficult to separate the land and the spirits from each other. The land is a living spirit.

[64] Van Onselen (1972) has a slightly different view. He argues that Africans blamed whites for maliciously doing something to trigger the rinderpest disease. According to Van Onselen (1972: 474), 'Rinderpest certainly formed part of the political backdrop to the 1896-97 revolt in Southern Rhodesia'.

There are different kinds of *chisi* that people are expected to observe. First, there is *chisi* day every Friday, when no one is allowed to go to the fields. Every new moon people are also expected to rest and *chisi* is also declared after the very first rains. While *chisi* itself cannot be considered a myth the punishments associated with any infraction are in themselves mystical because they are supposed to be meted out by the invisible hand of the ancestors. Punishment might not occur on the day of the infringement, but no matter how long it takes, 'as sure as the sun will rise the wrongdoer will be punished'; the popular proverb *chisi hachieri musi wacharimwa*. However, just like all knowledge, the issue of *chisi* is highly contested and various negotiations and renegotiations as to its significance have taken place over the years. As Bourdillon (1990: 263) notes, 'even sacred beliefs change in their force or precise meaning over time'.

According to the *Sabhuku* (headman) of Mudzinge village, rain is not just rain but it is composed of three kinds. Distinguishing between the different kinds is important since it could mean contravening a cultural taboo or not. The first rain after a long dry spell, or the one signalling the start of the season is called *Musana waamai* (mother's back), then there are *bumharutsva* (a rain of brief duration that occurs in August or September before the onset of the rain season) and *yemunhurukwa*. *Yemunhurukwa* is the rain that people use for planting their crops. The day after the first rains (*musana waamai*) no one is allowed to go to the fields because doing so will be breaking mother's back. In Shona culture breaking mother's back is taboo. For instance if a girl elopes or has a premarital sexual relationship with the man she will eventually marry, the husband-to-be will be expected to pay a fine for breaking mother's back. Even if the marriage is legal the son in-law still has to give the mother-in-law what some people refer to as *mbudzi yemusana* (the goat of mother's back) or *mbudzi yemushonga* (the goat of medicines and the medicines which are mixed with the meat for the back). Breaking mother's back for any reason is a serious offence for which a perpetrator is punished.

The headman explained that it was necessary to observe *musana waamai* because *vanhu vepasi* (people of the earth/ the lion spirits) will be inspecting the fields to ascertain whether the rain received was enough. If not then the lion spirits will bring more rain before people can start planting. Going to the fields on such a day would risk meeting *vanhu vepasi* and as a consequence one could be eaten by lions. However, *chisi* is heavily contested as people dislike it for a number of reasons. Most people except those in the chief's family regard *chisi* as a waste of valuable time. Farmers lose nine working days every month: four days are lost to the weekly *chisi*, a further four days if one is a Christian and has to attend church service any other day during the week and the ninth day is lost to the *chisi* of the new moon. All this is assuming that no death occurs in the village or in a nearby village, because this would entail the observance of yet another *chisi*, the *chisi* of death. Given the fact that the rains are erratic in the area, farmers maintain that they cannot afford to waste any rainy days, especially taking into consideration the fact that early planting could well prove to be their only salvation if the rains were to stop early. As a result, it is easy to understand the ire of farmers against the stipulation that they cannot go to the field on the day after the first rains. The situation is even worse when the rain is followed by a Thursday since this means that people cannot work on the Friday either. Below is a discussion about *chisi* I had with one family:

> *(Netsayi) Some people told me about* musana waamai, *what is it*?
> Father: That is *chisi*.
> Mother: That *chisi* pains us, I do not want to lie. When the rain has fallen and someone tells you that you are not supposed to go to the field because of *musana waamai*, it makes me mad. I think those people who are not able to farm and the lazy ones like *chisi* very much.
> Father: If you come here and do not get *sadza* will you visit us again? No. Then they tell us that when it rains we should not go to the field because of *musumo*. What is painful is this rule that people should not go to the field when it rains. We did know that is what they would make us do when we came here. They told us they were going to send us to the big farms (*kumapurazi*). We thought everything was going to be done the way we had done things at the commercial farms. We thought we were coming here to farm, not to do all these other things.
> Son: That's our tradition.
> Father: That is laziness.
> Son: If you don't do that, you will not get any rain. If you don't get the rain will you be happy?
> Father: Aah you! God is the one who brings rain not all these traditional rituals you practice. Rain belongs to God.
> Son: Are you saying traditional things do not exist?
> Father: Of course tradition exists (*chivanhu chiriko*). On the monthly *chisi*, people are not even able to eat *sadza* shouting that the new moon has risen (*mwedzi wagara* - literal translation, the moon has sat). Do you eat the moon?
>
> *What happens if you work your fields on* chisi*?*
> Son: You pay a fine.
> Father: God did not create us so that we would sit around doing nothing, no farming. He said humans had to sweat first before they could get anything to eat.
>
> *Is there anyone in particular who monitors the moon?*
> Son: The headman's policemen. It is important during the rainy season. Now people are not concerned with the moon.
> Father: It's because these people are lazy. Now because everyone is just sitting around not doing anything the moon suddenly becomes unimportant.

This conversation serves to highlight *chisi* as a domain that is now being highly contested because even people in the same household cannot come to an agreement over its worthiness. Those who are in favour of *chisi* cite morals and tradition as the basis for *chisi*, whilst those who are against it cite production reasons as to why the practice should be discontinued.

In the early days of resettlement, observation of *chisi* was not mandatory since the resettlement area was not ruled by chiefs but by a government appointed resettlement officer. However, since 2000 when the government started a drive to restore power to the chiefs, the resettlement areas have now been put under the jurisdiction of the chiefs. The first things that the chiefs did to make their authority felt was to reinstate *chisi* and to appoint an enforcement police that would patrol fields on *chisi* days and bring offenders to book. Most people in the sample said that the chiefs had done this purposely to 'fix' people in the resettlement area – a word which denotes a possible tension between the resettled people and the chiefs. The chiefs[65] had failed to understand the differences between the resettled and the communal people since the former regarded themselves as pseudo-commercial farmers because their main purpose was farming to feed the nation. They distinguished themselves from 'communal people' or 'people from the reserves' whom they said only farmed for subsistence because they lacked the requisite skills needed for commercial farming. Consequently, people in the resettlement areas were very adamant as to why they disliked *chisi*. They were not happy that they were being treated as communal people.

It is interesting that it was mostly the elders who opposed *chisi*. The youth, whom I would have expected to loathe the idea as old-fashioned, were at the forefront of those upholding it and argued that it worked and that 'our traditions' had to be upheld and restored. This goes against Fanon, who thought that youths would be the first to forfeit cultural norms and make fun of them:

> After centuries of unreality, after having wallowed in the most outlandish phantoms, at long last the native, gun in hand, stands face to face with the only forces which contend for his life - the forces of colonialism. And the youth of a colonial country, growing up in an atmosphere of shot and fire, may well make a mockery of and does not hesitate to pour scorn upon the zombies of his ancestors, the horses with two heads, the dead who rise again, and the djinns who rush into your body while you yawn (Fanon 1967: 45 cited in Lan 1985: xvi).

Instead the youths are at the forefront of defending custom.

On being asked what would happen if a person worked in his or her field on *chisi* days, one woman thought nothing much would happen. The only exception these days was that the rules were tightened so that if one was caught contravening *chisi* one would be sent to the chief's court for trial. The village police and the ZANU (PF) youths were responsible for arresting such

---

65 Read Roberts (2000: 513-22) on how certain narratives can be empowering for the people who propagate them.

people and sending them to the chief for punishment. Did the youth really believe in *chisi* or was this belief based on issues of power?[66] One youth believed that in a certain year when a drought was threatening, people were told not to go to the fields for four days whilst the traditional authorities went to the lion spirits to ask for rain. On the fourth day, there was a heavy downpour which confirmed that it was the work of the spirits. The father, however, was convinced the rain was an act of God; it is natural, either it is there or not, and nothing anyone would do could make it stop, or start, raining. I would like to posit that, especially considering *chisi,* a link between knowledge and power emerged so that those people who believed in *chisi* stood to gain a measure of power from it whilst those who did not were more often than not disempowered because of its enforcement. The youths could empower themselves by enforcing customary norms and enrolling everyone, even if unwillingly, into their own political and social schemes.

The way people have responded to *chisi* over time shows creativity and ingenuity. In the early days of the formation of the *Johanne Masowe* Apostolic Church, people attended church on Saturday as God had rested on the Sabbath day, so they too would rest and give thanks to the Lord. However, one day the church's very charismatic leader had a vision that the Lord wanted the church service to be held on Friday since the Lord had decided to anoint this day. As a result, the church became known as *Johanne Masowe wechishanu* (Johanne Masowe of Friday) to distinguish it from the *Johanne Masowe wemugovera* (Johanne Masowe of Saturday, which is located in other parts of the country especially in urban areas where people work from Monday to Friday). An unanswerable question of course is whether indeed the church's leader received a vision or whether it was an attempt to deal pragmatically with the *chisi* issue and minimise lost working days without confronting the traditional authorities.

At the level of farming people have responded well to new farming technology, such as winter ploughing, as a way of dealing with *musana waamai*. All people who practised winter ploughing in the resettlement area maintained that they had not done so prior to coming to the resettlement area. However, after being resettled and going into commercial farming, they could not waste any rains. Thus they ploughed their fields in winter, and then dry-planted their crops so that when the rains came and they were not allowed to go to the fields because of *chisi* their crops would still reap the benefit of the first rains. Tobacco farmers have also found a way around *chisi* as it is now the norm that people pick up a lot of tobacco on Thursday then spend the whole of Friday tying the tobacco in preparation for putting it

---

[66] Elsewhere in South Africa, Niehaus *et al.* (2001) discuss how the ANC youths were at the forefront of witch hunts in South Africa, and write about the possibilities for empowerment that such beliefs accorded these youths.

into barns. Thus the conditions that people find themselves in can make them receptive to new ideas as well as to devising new farming practises.

In proper observance of *chisi* no one was allowed to dig the earth since this was also the day when the *vanhu vepasi* (people of the earth) were resting. Digging the earth, therefore, was a sign of disrespect and would disturb the people of the earth from their well-deserved rest. However, as I would observe later, people would work in the gardens on *chisi* (a practice which most people were quick to point out was not allowed in the past). Yet, there was no agreement as to what exactly a person was allowed to do in the garden. Some people argued that one could only water the garden but not dig the earth whilst others maintained that people were allowed to do both. To explain this discrepancy the wife of the chief offered the very interesting comment that sometimes things are discussed at the village court and new pronouncements, which override old ones, are announced. However, some people hold on to what the spirits said long ago and thus do not keep track of any new pronouncements. Accordingly, many things have changed and, as said, clever chiefs take advantage of the time after the spirit medium has died and before a new medium comes out, to make any changes they want. In most cases, after such changes were made, later mediums would rarely dispute them because it was believed that when the medium was not there, the chief could only make inspired decisions, unless of course the chief was unpopular with his people.

However, early changes to *chisi* were made before the spirits became silent. People were not allowed to go to the fields because the *vakuru* (elders, again referring to lion spirits) would be inspecting the fields in their lion form. People reasoned that the gardens were fenced and as a result *vakuru* could not get into the gardens. Hence, their walk would not be disturbed by people working their gardens. People also asked for permission to water their gardens on the pretext that in the gardens they would use *mvura yechirimo* (dry season water) whose *chisi* had already been observed during the previous season. When asked if the same would apply for irrigable agricultural land, people were adamant that it would not work since the area was often too big and there was no guarantee that whatever was being irrigated would not receive the very early rains of the next rainy season. On 10 January 2003, a new pronouncement was made by the *sabhuku* in Muringamombe: people could now weed their field crops provided they were planted in their back yards.[67] It was argued that the lions only roam the

---

[67] Resettlement officers had prohibited people from growing crops around their homesteads as this was thought to cause diseases like malaria. Homesteads also had to be kept clear of agricultural crops since thieves could hide in them, waiting for an opportunity to strike. However, just like people resisted the edict to stop them from keeping goats or to stop stream bank gardens, they also resisted this move and backyard farming became common. Also see Spierenburg

fields but do not come inside the villages where people live. It was therefore on such a *chisi* day that I found my assistant and myself weeding a back yard tobacco crop.

In the same village there was an outright resistance to *chisi* and I had the fortune to see some people working in their fields on *chisi*. People said they were probably working for *maricho* (payment) and presumably this was less serious than to be working in one's own field on that day. Surprisingly, the people were not arrested even though they were working near the roadside and in full view of anyone who was passing by. Given these varying applications of the *chisi* taboo, it turns that people are not just docile, accepting things simply because they come from authorities, but find alternative ways of manoeuvring and in the end doing what they want to do.

People sometimes respected *chisi* not out of fear of the ancestor's wrath but because of practical consideration, especially where *chisi* of death was concerned. If a person died in the village or the neighbouring village, people were not expected to work their fields but to attend the funeral. Most people, even those who resided at their fields, claimed they respected the *chisi* of death, as they had to assist fellow villagers to deal with their bereavement. At funerals people helped with cash contributions, food contributions, with cooking, consoling the bereaved family and digging the grave. Support from fellow villagers during bereavement is very essential. People do observe the *chisi* of death, not only because it is mandatory but because there is a genuine fear that if you do not people will not come to your aid should such a misfortune befall your family.

However, some things have been amended as people are allowed to tie tobacco at home if they reaped it before they knew there was a funeral. Thus although in the past people were not allowed to do any agricultural work on *chisi* except herd cattle, with the advent of new crops such as tobacco, which could easily go bad if proper procedures are not followed, the rules are changing. Most farmers did not believe that *chisi* was helpful at all but observed it because they did not want to antagonise traditional authorities.

In the past it was believed that if a person worked on *chisi* his/her field could be attacked by marauding animals sent by ancestors, or else the person would see terrifying things like a monkey or baboon dancing to a radio, or a python coiled around his plough. In extreme cases the perpetrator could be struck by lightning. These things are no longer seen to be happening and some people were complaining that they did observe *chisi* yet marauding animals often attacked their field instead of the fields of those who were not observing *chisi*. *Chisi* is no longer self-enforcing, and people can make their interpretation of *chisi* suit their own ends. The people who still cling to the

---

(2004) on how people in Dande, as well as some project managers from development agenicies, negotiated with the lion spirits for their projects to be allowed.

traditional idea of *chisi* are the chiefs and the *sabhukus*, and others like the youths who stand to gain if the traditional structures are kept intact. However, the two *sabhukus* were unfortunate in that they were unpopular. One was seen to support *chisi* because he was very poor and very lazy, and the other one because he was crazy as he had just recovered from a mental illness caused by witches.

Still with regard to *chisi,* events in the village can be interpreted differently by different people to suit their own agendas. One such incident occurred in the early 1980s when three people in the village lost their cattle to marauding lions. One individual lost two cattle and the other two individuals lost a cow and an ox each. All the people in the sample agreed that when they first came the area had thick trees and forests, which were a habitat for many wild animals including lions. I heard three different versions of this story, depending on the people I talked to.

One of the villagers mentioned that Nyamaropa (the lion spirit of the area) had done it because the man who had lost two cattle had worked on *chisi*. Suffice to say, the man was being punished for his sins. However, the respondent conveniently failed to mention that others in the village had also lost their cattle on that same day. This is selective retrieval to suit whatever point people want to put across. The second respondent, who was among those who had lost cattle on that fateful day, did not think it was because of *chisi* violations, because she had never worked on *chisi*. However, she thought it was not a mystery, given the number of lions that roamed the area at that time. The third respondent was convinced that this was a clear case of punishment from the spirits:

> Chisi hachieri musi wacharimwa (*one does not get punished on the day that one breaks the chisi taboo*). One year we experienced a drought. My brother Josiah, the apostle, took water from the river fetching it with drums and a scotch cart to water his field. The spirits did not like that so his cattle *Mazai* and *Misisi* were attacked and killed by lions. Children who were going to school saw the cattle lying in the road in a strange way and they quickly came back to report. *Madzisahwira* began to ask him how he could water his tobacco using water from the river. Now look at how the spirits had retaliated.

In this incident we could witness three different understandings of the same event. The first one is yes, there were a lot of lions at that time but this was no ordinary lion but Nyamaropa who was punishing a violation; (2) there were a lot of lions and the lions would sometimes attack livestock; (3) Yes, there were a lot of lions but this was Nyamaropa's work; however, the person was being punished for irrigating his crop using river water. In this case, three different knowledge assertions were generated. These assertions converge at some point and diverge also on some fundamental points and each one is convinced that what they are saying is the truth for which they have evidence.

Christianity

The Christians did not believe that the *zumba* could ensure seed fertility, or rain. However, they did contribute to the seed that was taken to the *zumba* as well as to the grain that was needed for the rain-making ceremony. As one respondent who did not believe in the *zumba* commented, 'when you are in Rome, do what the Romans do'.

Christians mostly point out that rain and fertility come from God, not from some spirits of the land. At the start of the season they would take their seed to the various priests of their various denominations for prayers to be blessed. The seed would then be mixed with the rest of the seed at home before planting. This would ensure that the crops grew healthy and strong and it guarded them against *tsvera*. Any kind of seed, treated or not, modern or traditional, could be taken to the priests. For the Christians, there was no distinction between modern seed and traditional seed in terms of performing fertility rituals. The apostles went further and would get holy water to sprinkle onto their fields or bury bottles of water in their fields to guard against *tsvera*.

Apostles

Although the apostolic churches operating in Mupfurudzi differ in a number of ways, they are fundamentally similar. For example, unlike the missionary churches the apostles emphasise speaking in tongues, seeing visions and having prophetic dreams. Thus, during disputes dreams and visions can be used to support particular positions. Through praying, prophesying and fasting, believers can be protected from ill-health and poverty. Daneel (1987: 235) correctly points out that for the apostle there is a clear causal connection between human activity and the absence of rain. For example, both Marange and Masowe preached that if people did not do as God commanded them, then there would be drought and diseases. If people desist from unholy deeds like witchcraft and adultery then God will bless everyone with abundant rain and bumper crops.

I had a conversation with one woman who was a member of a very small and unknown apostolic church called Borngesi. This is what she had to say:

> *We once discussed traditional beliefs. You said you do not believe that stuff. However, I want to know whether you send your seed to the* zumba.
>
> We just give them the seed. They will be doing it for the rain so even if we do not want we will just end up giving them our seed. After all, we will not use the seed in our fields so there is no harm done. Someone refused to vote during the elections and he was told to live in his own world where people do not vote. That is what we are afraid of. At the end we just do everything we are asked to do,

> even though we might not believe it works ... We still have to send the seed to the *zumba* because when it rains can I say it should not rain at my field because I am a Christian and the rain is from *zumba.* I would be the first one to take my plough to the field regardless of where the rain comes from.

For abundant rain the apostles also climb the mountains to pray to God who will hear their prayers just as he hears and answers their prayers for the sick. Thus, at the beginning of the rainy season, when the elders go to the *Zumba* in the wilderness to ask for rain the apostles also go to ask for rain by praying hard, fasting and preaching until God gives them a vision of what is going to happen. They will receive word on whether their prayers have been answered through the various prophetic messages. As a result, there is confusion among some people, as shown in the above excerpt, as to whether rain comes from the Christian God or the *zumba.* The confusion is not only found among the Christians but also among the traditional authorities. When one village headman was asked about rain-making ceremonies, he said that sometimes he even went to Wimbow to ask for rain at the people's request. Sometimes non-Christians use the behaviour of the apostles to validate their own behaviour, and likewise the apostles look towards tradition to validate theirs. Asked what she thought of taking seed to the *zumba,* one woman replied: 'That is the law. Even the apostles take their seed to Wimbow so that he can bless the seed. Wimbow then distributes the seed back to the people. It is the same thing with our culture'.

While people adhering to African religion take their seed to the *zumba* for fertility guarantees, so the apostles take their seed to their prophets for fertility prayers. (However, note that not all Apostles believe in the efficacy of these apostolic fertility rituals. One Johanne Marange Apostle was adamant that only the proper use of fertilisers could guarantee good harvests and not holy water from priests.)

Also, just like the Lion Spirits in African religion, the Christian God of the apostles is concerned with the general well-being of his followers.

> I still remember one drought year when Wimbow asked people from his church to come to him so that he could reduce the size of their stomachs through prayer. This was so as to make sure that they did not feel the need to consume a lot of food.
>
> *Was there any change?*
> What change? It did not work. These people. When it came to eating we sometimes left them in the plate, meaning that their stomachs were as big as ours were. Maybe ours were even smaller than theirs (laughs).

> *Since this is a drought year have people gone yet to have their stomachs reduced in size?*
> This year he has not yet said anything. Maybe they will go later because the hunger has not really started. Maybe the church members have already been informed about what they are going to do.

I talked to people, some of whom were not members of Johanne Masowe,[68] who claimed that Wimbow had indeed had their stomachs reduced in size and they had not starved during the drought year because of that.

Discussion

In the foregoing section, I sketched some of the main features of the two religions dominant in the area. Now I will discuss how these two dominant religions impact on knowledge and knowledge production. It is not religion *per se* that determines what a person chooses to believe and what is relevant knowledge. Sometimes it is the social positioning of the individual that is crucial. Knowledge is a result of situated selections. For example, one poor farmer who was also the village head strongly believed in traditional customs, firstly as the custodian of local culture and most importantly because it gave him an element of power. Because he was the person who could communicate with the ancestors, people gave him respect which he would not have gotten if he had not held a traditional office. The same goes for the ZANU (PF) youths who strongly believed in the power of the ancestors even though their parents might have been of a different persuasion. These youths gained position in society by enforcing traditional regulations and punishing offenders. On the other hand, those who can gain neither power nor office by following the traditional rules, at certain times believed that, regardless of what the spirits of the land thought, good farming methods would ensure good yields, whilst at other times they would point to the importance of *mhondoro* or *midzimu*, or God.

The tension between local knowledge and western knowledge is epitomised in the ritual of the *zumba*, where no modern seeds are allowed. The *zumba* constitutes a ritual rejection of modernity and modern farming methods, which are perceived to harm the land, and a reaffirmation that the ancestors are concerned with the well-being of their people. This they do by focusing on food crops to ensure food security. However, this does not mean that those who believe in the power of the *zumba* do not cultivate modern crops. On the other hand, people who do not have access to certain resources can use the *zumba* or other beliefs attributed to Shona religion as a rationali-

68 Wimbow now refers to his church more and more as *Vadzidzi* (The Disciples). I do not know whether this signifies a break from mainstream Johanne Masowe.

sation for not doing certain things, regardless of whether or not they believe in Shona Religion. For example, in the army worm outbreak of 1985, those who had cotton chemicals sprayed their maize. Those who did not spray claimed that it was because they were afraid to anger the owners of the land since they had not been given permission to do so. At least one of the people who cited respect for the owners of the land as a reason for not spraying their crops was a member of an apostolic church that deplored anything that smacked of African spirits and ancestors. Appeals to African spirituality could have been a legitimate reason for not spraying the crops but this does not mean that it was the only or main reason.

Christians, and those who subscribe to Shona religion, are not very different in their farming beliefs. For them some power determines fertility, rain and good fortune. No amount of fertiliser application or even good rains could save you from the wrath of the Gods (whether African or Christian). If the Gods are displeased no matter how hard you have tried you will still lose out. This is where the knowledge of agricultural experts and people differs. For the experts, agriculture entails proper planning, a planning which starts with buying the requisite seed and fertilisers on time as well as preparing the land on time. For the people, good agricultural knowledge also entails proper planning, a planning which does not start with seed and fertilisers but with having a good relationship with the Gods and ancestors. Only after you have set up a proper relationship with your God and ancestors will you be satisfied that your seed and fertiliser have power. Nevertheless, a perceived interdependence exists between religion and farming technologies. For example, people realise that one cannot simply rely on religion without getting the necessary farming implements, and farming implements without proper religious backing will not take you far either.

Indeed the apostolic faith churches borrowed much from Shona religion, which might explain their popularity among the people. Burying bottles of water which had been blessed by priests to protect one's fields was not much different from digging in clay pots filled with water and medicines to also protect one's field. The only difference lied in the interpretations accorded to each act. The clay pot was regarded as a sign of *tsvera* and not protection, while the bottle of the apostle could never be regarded as *tsvera* but only as protection. Blessing the seed or taking the seed to the *Zumba* can both be regarded as *divisi* since both actions are meant to ensure good healthy crops. However, they are both good *divisi* since they do not encourage the breaking of incest taboos. Taking a cue from Aschwanden (1989), the virgin sisters of Johanne Masowe, who were dedicated as the wives of God through whom all requests should be passed on to God, could be regarded as the ultimate sacrifice. That is, a sacrifice of fertility so that God could hear man's requests whether they be for rain or the fertility of crops. As mentioned earlier some 'witches' who used *tsvera* or other kinds of *divisi* also sacrificed the fertility of their daughter(s) by dedicating them as wives of goblins and *Chikwambo* so that they never married. If the goblin was kept happy then the *divisi* would work.

For most people, religion has a direct influence on their ways of thinking and evaluating data available to them. If a person strongly believes in something then that belief may preclude all other possible explanations of events. People and groups do not necessarily adopt any kind of information they receive from outside, since they retrieve information and act on it, according to their perceptions. These perceptions could be influenced by religious beliefs as well as by what is seen to be beneficial. For example, people mentioned that when they first settled in the area there were a lot of trees and animals and the rivers were overflowing with water. However, since then people have cut down most of the trees, cultivated gardens along riverbanks (violating the rules against stream-bank cultivation) and have panned for gold in Mupfurudzi river. Nowadays the animals are nowhere to be seen and the rivers are drying up. The explanation for the rivers drying up and the animals going away, as mentioned earlier, was that people had started using soap in the river, which was against the dictates of the lion spirit. This had angered the lion spirit, who dried the rivers. An alternative explanation could well be that the rivers are silting up because of gold-panning and stream bank cultivation.

Linkages also exist between local religions and the Christian church. For example, a local traditional chief who led the traditional rain-making and fertility ceremonies was a member of the Salvation Army church. This was not considered conflictuous, as the Chief's wife argued that the bible said *to Caesar what is Caesar's*. Also most people drew parallels between Christianity and African religions to justify their positions. For example, concerning seed fertility ceremonies people would argue that they had to do this because it was the law and even the Christians took their seeds to their priests. Thus, Christianity and African religion reinforce each other in the face of doubters like the agricultural scientists.

Another important consideration is how gender difference can impact on religion and knowledge. Invariably the women tended to be more religious than their male counterparts and mostly cited their religion as Christian. As mentioned before, it was more usual for women rather than for men to take seed to the priests or go to the priests for holy water. On the other hand, apart from when the women brewed beer, men were mostly in control of the traditional ceremonies. Although women could be mediums of powerful spirits, the role of women in traditional ceremonies was generally marginal. Women took part in brewing the beer for the ceremony but they were not involved when the major decisions were made. Major decisions could only be made by the chiefs and village headmen (who were all men) and the rainmaker (who sometimes could be a woman possessed by some powerful spirit). Where men expressed some scepticism in the ability of the traditional ceremonies to ensure bumper harvests, all women were adamant that their holy water could ensure a healthy crop and guard their crops against bad magic. As mentioned earlier, it was usually the men who were in control of obtaining all farming implements. Women did not have control over those things, though at the same time they were expected to ensure food security or else risk going to *maricho*. This could explain why more women than men turned to religion.

Conclusion

Knowledge is largely interpretative in nature. Knowledge from the same sources can be interpreted differently by different people. This means that knowledge itself is not an entity out there waiting to be used but is socialised and re-socialised in different ways by different people. Interpretation is locally specific and contextual. In the end, people can understand the same phenomenon differently, giving rise to different knowledge claims. Even people from the same household, as indicated in this chapter, can sometimes have conflicting views on the existence and efficacy of magic in enhancing farming ability, leading household members to explain the same phenomenon differently. In such circumstances, people or household members might call upon different bodies of knowledge to explain certain phenomena that structurally look the same. As a result, rural development workers should not be overly scientific, ignoring the different perceptions and meanings that people attach to certain activities and interventions, as this could spell the failure of scientifically sound projects.

Belief in magic does not directly impact on agricultural production. Rather, magical beliefs are used to explain failures, but when people believed in themselves this did not constitute a reason for failures. People did not rely on magic to get good crops but also applied fertilisers and bought good seed. Even those who used holy water from the church or took seed to the *zumba* still used other technologies to increase their yield. Only those who did not have access to enough fertiliser and good seed to get good crops often blamed their low yield on magic. Hence, religious and cultural beliefs should not be regarded as impediments to increased agricultural production, because such beliefs do not preclude the adoption of more productive methods of farming. The tension between African beliefs and modernity is largely imagined. Discussing ritual powers among the Tuareg, Rassmussen (2004: 318) notes that

> ... these powers do not imply neat temporal oppositions between 'tradition' and 'modernity', or linear regression from one to another … these powers suggest plural and interweaving, rather than singular or sequential, moral discourses of tradition and modernity.

In this chapter, the interweaving of outside and local knowledge and diverse local theoretical traditions has led to knowledge that defies attempts to be strait-jacketed into categories of traditional or scientific.

There is a mistaken belief that African beliefs are static and resistant to change and that this rigidity of African beliefs has proved to be an impediment to agriculture. On the contrary, as shown in this chapter, African religion and culture are highly flexible and mechanisms exist within it for generating change. This ability to change has enabled African religion to maintain its relevance even in the face of the onslaught of modernism. Instead of looking at beliefs and culture to explain why people refuse to adopt certain technologies, experts should also begin to question the technologies themselves.

One needs to identify the cultural dialectics that make it possible for people to adopt new things that are of use to them. Local forms of knowledge are always being reworked in interaction with changing external and internal conditions. Thus local knowledge should not always be regarded as resistant to change. People could not control *tsvera* effectively and therefore were always afraid of losing their food security. The use of fertiliser and other effective farming methods was, as it were, another kind of *tsvera* but one they could take full control over, thus guaranteeing food security providing they followed its rules.

Since *tsvera* is only associated with food crops, not cash crops, AREX and other outside agencies could focus on the latter. As discussed in previous chapters, farmers can import what they know from the cultivation of cotton and tobacco to other crops, since the basic principles of farming are the same for all crops. All crops require that one prepares land early, obtains good seed and enough fertiliser and plants on time. Therefore, good cash crop farming could filter down and to improve the cultivation of food crops.

Shona religion gave rise to certain knowledge. This knowledge functioned within the system, assigning power to some people within the system but not to others. The same can also be said of modern knowledge and technology that gives power to government bureaucrats. Knowledge assigns power within the system, resulting in functionaries such as the village headman and ZANU (PF) youth clinging to some African beliefs that give them certain powers, and trying to change those things within African religion that do not suit their projects. On the other hand, ordinary people and farmers expressed their desire to move away from these systems which they saw as inherently disempowering, although they also clung to traditional beliefs and knowledge when it suited their needs. Thus, knowledge is partial, indeterminate and sometimes self-contradictory.

In contrast to the common belief in popular literature that the peasant's mind is difficult to change, local farmers always negotiate and renegotiate their knowledge and beliefs to suit current needs that give rise to new world views. The point has been explored in this chapter in relation to the *chisi* practices among the Shona. People have found novel ways of modifying and contesting the enforcement of *chisi*. 'Traditional' knowledge is therefore not as traditional as it may seem at first sight, as people 'always process social experience and ... devise ways of coping with life even under the most extreme forces of coercion' (Long and van der Ploeg 1994: 66). However, farmers can only change their behaviour and attitudes if the required changes are within their means and are perceived to be beneficial in both the short and long term. Development experts should first understand farmers' needs before they are quick to condemn them for being resistant to change.

6

Field days: Knowledge dissemination and entertainment

Introduction

Field days were introduced as a way of involving the farmer into the modernisation agendas of the 'experts'. 'By means of celebrating agricultural success demonstrated in the fields ... it was hoped to induce less successful farmers to copy or mimic prize winning Master Farmers' (Bolding 2004: 95). In line with the Transfer of Technology approaches, officials understood field days as occasions for AREX officers to impart agricultural knowledge to farmers, thereby assuming that farmers lacked such knowledge and needed to be tutored (Hedrick 1918). Unlike Master Farmer classes (discussed in Chapter 1), field days emphasised the practical rather than the theoretical side of knowledge, or knowledge disseminated in the context of the classroom.

Field days were also understood as occasions where farmers could share knowledge with each other in a guided and controlled environment. Although the host farmer was given the opportunity to explain to others what he did to get a good crop, the expert was present to offer guidance and correction when farmers erred in their explanations. This contradicts Bolding's (2004: 100) assertion that

> ... most field workers perceive field days not as learning events fostering the effect of trickle down otherwise known as extension but events to legitimise their *raison d'être* as state agents committed to modernity, development and commercialised small holder agriculture (see also Worby 1995, on the ritualistic nature of field days organised by Agritex in North-Western Zimbabwe).

The data in this chapter will show that, although there might be many reasons for holding and attending field days, officers do take seriously their teaching and learning functions.

Photo 6.1 Mr Kadungure, the host of the field day in Magazi, explaining to other farmers with his two wives standing by his side. Photo by Ellen Luka (field assistant).

Officers disseminate knowledge about seed through pre-planting meetings and field days right through the season. Field days give officers the opportunity to disseminate information in a relaxed atmosphere, different from the school-room atmosphere. Such days are usually entertaining, with people performing dramas and singing songs of a political or agricultural nature, and in some cases there is an abundance of food and beer, though not at any of the field days I attended. All farmers get a chance to participate and share their knowledge.

Photo 6.2 A women's choir club at a field day. Photo by author.

Apart from being situations to share and impart knowledge, field days are also occasions where social hierarchies are recognised, reinforced and disputed. Although designed as occasions to discuss production issues, they are also events where dance and drama are used, to highlight farmers' problems as well as to entertain. Some people might also take a field day as an opportunity to solve long-standing disputes and to gossip. In some instances field days are highly political. I will start the discussion by focusing on how agricultural knowledge is spread at field days.

Field days and agricultural knowledge

Both farmers and experts took field days as learning occasions. However, as will emerge in the course of the discussion, the emphasis that farmers and field officers put on field days as learning occasions differed. The AREX officer and the Cottco Collection Point Supervisor mentioned field days as the special days they set aside to talk to farmers. Farmers, however, were disappointed that they only met field officers at field days and other large gatherings. To understand the impact of field days as instruments of agricultural knowledge dissemination, it is important to investigate these differing perceptions.

Farmers and Field Officers operated from different premises. Farmers saw the officers as having been mandated by the government to work with them on a personal basis, as they had done in the early days of resettlement.

The failure of officers to offer this personalised service was regarded by farmers as a betrayal. Furthermore, some farmers were too timid to ask questions at large gatherings for fear of revealing their ignorance. One poor household head and his wife pointed out that they were not very happy with the conduct of the AREX Officers. No AREX officer had ever visited them to give them advice. They accused the AREX officers of sitting in their offices the whole day doing nothing, instead of going around inspecting people's fields and giving advice where appropriate. They claimed the officers of the early resettlement days used to do their job properly, but not the current crop of officers.

On the other hand, especially where AREX is concerned, the government has embarked on internal restructuring, sending some of the officers to the fast track resettlement schemes. At the same time there has been a noticeable drop in the number of extension workers due to the restructuring of the civil service. For instance, Mutangadura (1997: 37) writes of an AGRITEX with a staff component of 2,500, whilst Murwira *et al.* (2001: 302) write of an AGRITEX with a staff component of 2,000. Also, with the dismantling of AGRITEX to form various specialist departments, of which AREX is one, the number of extension workers focusing on crop production went down. Instead of six AREX officers operating in Mupfurudzi, there was now only one. A single officer could not deal on a personal basis with all the farmers, and therefore could only meet them at field days where there was usually a large gathering. Farmers would be encouraged by the officer to ask questions. Another villager who was not pleased with the way the AGRITEX was being structured said:

> They only come to inspect *madhunduru* (contour ridges). What they had told us was that they were going to have an AGRITEX officer concentrating on tobacco only and another concentrating on the rest of the crops. They were supposed to operate from Zvomanyanga to Chidubwe near Bindura. That has not yet happened but I guess it would not have worked out. One AGRITEX officer operating in such a large area. It is better for them to tell us that they are no longer going to give us AGRITEX Officers. How can one officer be expected to cover such a large area effectively? In the early days they used to tell us to educate our children because they said that with time AGRITEX officers and Resettlement Officers were going to be phased out and we would be left to run our own affairs. I think that is what is happening now. One can ask questions at village meetings but then sometimes people will laugh at you after the meeting. Like that man over there who used to tell us that he was the only good farmer in the village each time a field day was held at his place.

AREX officers and the representatives of the company sponsoring the field days took it upon themselves to choose the 'good' farmers to host the field day. As mentioned in earlier chapters, the field officers and the farmers viewed the concept of good farmer differently. The Cottco Collection Point

Supervisor (CCPS) made it clear that Cottco was concerned only with the production capacity of the farmer:

> We assess whether the crop in the field is healthy. We also look at weed management and pest control as well as the amount of feeding that he gives his crop. We look for someone who performs better than all the other farmers. If he gets more than 30 balls per plant, then judging by the standards used here, the person will be good.

Whilst the field officers considered only production factors, farmers also considered things like suspicions of the use of magic, as well as how the farmer related to others. This could determine whether people were willing to attend a particular field day or not. One farmer mentioned that he did not like attending field days held at a particular man's field because the man was arrogant and boastful and thought he was the only farmer around. Three farmers in the sample were disappointed that the field officers sometimes did not hold the field days at the real good farmers' fields but instead chose farmers who farmed with the help of bad magic. Field days in this way failed to achieve their goals because officers did not have a full understanding of the local politics. Knowledge is political. Officers were also excluded from village gossip: not many people were willing to tell them anything since they were sometimes believed to be good friends of those who were suspected of using magic.

Field officers regarded field days as a way of encouraging good farming practices by fostering envy and the spirit of competitiveness among farmers. Extension officers and private companies tried to create competitiveness by giving farm inputs as prizes for the farmer who was judged good enough to host the field day. It was assumed that other farmers would also want to get free inputs and gain recognition as good farmers and would thus try to improve their farming. For example, a farmer selected to host the day would get inputs from the Cotton Company sufficient for one hectare. At the national level, a small farmer selected as the grower of the year would receive a million Zimbabwean dollars (equivalent to US$18,181 on the official markets and US$1,166 on the illegal parallel market at the time), inputs, a scotch cart, a water cart, a rain gauge and a T-shirt. Apart from the material rewards on field days, good farmers were often praised by the experts and the villagers would praise the farmer in song, which would make the farmer feel good. For example, at a field day hosted by Mr Kadungure, people sang:

> We are proud of Mr Kadungure
> He cultivated cotton that has a lot of balls
> He will get a very good yield
> Come and observe so that next year you farm

We are proud of Cottco: it gives us loans
To advance the farmer
We are proud of Mr Mushayi and Mr Jonga,
They give us knowledge so that we can progress.

At times AREX would choose a few farmers and give them seed to cultivate on demonstration plots to be observed at field days. Why these farmers were selected was not made clear to the farmers. As a result, some farmers pointed to favouritism. However, the AREX officer declared that sometimes for demonstration plots, instead of choosing a 'good farmer' they would select a farmer who could obtain fertiliser and would agree to close monitoring by AREX on the demonstration plot throughout the season. As a result, those who did not like the idea of close monitoring were left out even though they were good farmers and could afford fertilisers. Some farmers were known to be short-tempered and did not take kindly to being told what to do.

Although field days can be regarded as a viable option of spreading information when alternatives are restrained, they are still limited in their scope. Instead of being learning experiences, to some extent they become more like road shows for companies to advertise their products. For example, if Cottco sponsors a field day, no questions can be asked on any other variety that is not produced by Cottco. As a result farmers that do not cultivate Cottco seed or deal with other cotton companies, like Cargill, get no clarification of their problems, unless their company also organises a field day.

During field days, people sang songs praising the company that was sponsoring the field day and often ridiculed other companies. The following are examples of two such songs that were sung by Magazi and Chiedza Women's Clubs respectively, at a field day in Magazi:

All of you get lost
We love Cottco because it is good
Agricom get lost
Tsikamutanda get lost
All political parties get lost
We only want Cottco it is the only one
All of you get lost
Cottco is very good.

Chiedza club women sang:

We have our seed, holding our weapon
We are going to unseat Cargill
We have Mr Mushayi we are going to unseat Cargill
We have Mr Jonga we are going to unseat Cargill

We have Mr Kahari we are going to unseat Cargill
We do not fear Agricom we are going to unseat Cargill.

Such field day songs suggest that ostensibly, field days are a good occasion to advertise the company that has sponsored that particular field day at the expense of its competitors.

According to the CCPS, Cargill is Cottco's biggest competitor in buying cotton from farmers, with Agricom in third place but not as a very significant player. Tsikamutanda was a local name given to a Tanzanian company that also purchased cotton. Tsikamutanda is a name given to some witch-hunters who sniff out witches by divination and mysterious means. Applied to the Tanzanian company, the name could have been derogatory, as these witch-hunters do not get favourable media coverage. However, it could simply point to the mystery surrounding this company as little was known about it except that it came from Tanzania, a country that also was mysterious to most of the villagers. Unlike the powerful Cargill Company that needed 'unseating', Tsikamutanda and Agricom could just be pushed over.

Although the farmers composed these songs it did not mean that they believed in what they sang. If another company held a field day (usually Pannar Seed and Seed Co - maize seed houses - and Cargill, which specialised in both maize and cotton) the same groups might perform the same songs and simply swap the names to suit the company sponsoring the day. During the performance people would mention and praise the company officials and field officers, who would then feel obliged to give money to the performers, a practice, popularly known in Shona culture as *kupfupa*. These performances could therefore be regarded as clever fund-raising initiatives by the performers. Laughter could often be heard from the audience who would be saying, 'this group knows how to shake the tree' (make the officials give them money) or, 'that group does not know how to shake the tree'. Officials usually had prior knowledge of the number of such groups so they brought with them enough money to give to them maybe as a token of appreciation of their performance or just to fulfil an obligation.

All performances whether song or drama, had to be related to agriculture. No song or drama could be performed just for its entertainment value. Those who could neither sing nor act can dance or clap hands.

Even primary school children who took part in the entertainment by singing or dancing to traditional songs selected songs that had something to do with farming. The following are examples of the songs they sang. The first song shows an appreciation of the good work that farmers do to keep people well-fed and nourished:

I thank you, grandmother, I thank you
Thank the farmer
I am full with food, grandmother, I am full with food

Thank the farmer
Thank you, grandmother, I am full with food
Thank the farmer.

The second song pointed to the consequences of being lazy. Lazy people led miserable lives:

This year I am not going to farm
Heha he-e
This year I will eat goat dung
Climb on top of the mortar and tell those at home.

Thus, field days can also be regarded as festivities to celebrate the farmer.

Some of the dramas discuss issues relating to lack of access to loans or unfair treatment by loan officers, or of farmers who lose their crops to unscrupulous salesmen who pose as buyers. They serve as a way for farmers to inform officials of their concerns and problems. In their speeches the officers frequently refer to some of the issues raised during the performances and they sometimes provide clarification, a solution or a promise of a solution after consultations with their bosses.

Why people attended field days

People had a variety of reasons for attending or staying away from field days. Of eight people asked, only two attended field days regularly. They maintained that they attended because they could get knowledge and information from other farmers and the experts. Some farmers attended field days not only as a quest for knowledge but also to escape stereotyping. For instance, not attending can sometimes carry negative connotations such as being labelled lazy and or ignorant. These stereotypes are not mere words but can have real consequences. The companies that sponsored field days also offered loans. If the sponsors thought you did not attend because you were lazy or resistant to acquiring farming knowledge and improving your farming skills, this could reduce your credit worthiness and sometimes even the ability to secure anything on loan.

Others attended field days to maintain good relations with fellow villagers. Most people, especially women, only attended field days when they were held within their villages. Although one woman felt that people could learn a lot of useful things, she admitted that many women (herself included) usually did not attend if they were held in other villages because women were too lazy to walk long distances or maybe frail health prevented them from walking such long journeys (one woman cited a heart condition). Another reason why women do not attend could be that they are busy attending to their domestic chores. Often times distance from home was mentioned as a huge factor militating against women's attendance of field days. It could be that, apart from the singing and dancing, women felt that, they did not have anything significant to contribute and felt that whether they attended or not, the field day would not be affected in any way by their absence. One woman only attended the field day that was held in her village. If she had not done so, people would have viewed her with suspicion or even have openly accused her of being jealous of other people's success.

One woman said she had never attended a field day, because she felt only men were encouraged to attend, as household heads, to learn about farming, so that they could successfully feed their families. When I pointed out that at the field days I had attended there was a large contingent of women as well, she simply pointed out that, women also attended but mostly to cook for the people participating at the field day. This was true: most of the women I saw at the field days were there as entertainers. Apart from the government female workers, who did not take part in entertaining the audience, all the women were involved either in the dramas, or in cooking or singing in choirs. The women took pride in their singing and dramas and it is highly probable that women mostly attended the field days so that they could perform. As mentioned earlier, they were given money for performing and they usually did not take much interest in the official proceedings as they were constantly reminded to listen instead of making noise.

The reluctance of women to attend was not limited to field days but extended to other meetings. The AREX officer felt that at other meetings women and men did not attend proportionately.

But it is possible that the high attendance of women at field days compared to other agricultural meetings was because field days had high entertainment value and offered opportunities for women to make money as well as point to their roles as cooks. At other meetings no food was cooked or offered but on field days there was always food and women had to be there to cook it.

Only one woman said she did not attend field days when her husband was alive, because he had been a jealous husband and had forbidden all his wives from attending any such occasion. Consequently, even after his death she never really felt the need to attend any field day.

Some men also did not attend field days but the reasons for men were different from those of women because of the different demands on their time. As mentioned earlier women had plenty of domestic chores and also cited distance to the field day venue as a militating factor against attending field days. The longer the distance the less likely they were to attend. For men distance was usually not an issue. One man said that he had not attended for the past two years because he often did not get to hear about them. Usually he got the news when the event had already taken place. This could have been true because on two separate occasions I faced difficulty in trying to find my way to the places where the field days were being held. When I tried to ask for directions, villagers claimed that they were not aware of any field day being held in their own or even in the neighbouring village. One young man maintained that he did not attend because he did not like the fact that he might have to spend the whole day hungry. He expressed nostalgia for the past: 'In the past field days used to be a place of feasting, but now those feasts are no more'.

He claimed that in the past people would drink beer at such events and a cow/ox would be slaughtered for people to eat. In the past, he had usually attended these field days to feast but now all that was gone. This young man liked drinking and was often to be found at the local bottle store.

Interaction at field days

Field days were an opportunity for the host farmers to explain to other farmers how they managed their farming activities. This was usually done during a tour of the field. The tour usually consisted of going to that part of the field where the crop was greener or more fruitful. The farmer would stand in the crops and explain to the other farmers standing in the walkway what he did to achieve such a good crop. The farmer was allowed to explain without interruption and farmers could then ask their questions. In the explanation the farmer included information about the days that he had planted his crops, how he planted them, what kind and amounts of fertilisers he had used and the kind of seed. In this instance the farmer would be the male head of the household and the women were usually forced by public demand to stand beside their husbands while they explained the intricacies of farming. Farmers might also discuss any obstacle they faced, and at the field days I attended, farmers often pointed to the lack of access to fertilisers. First, the farmer was asked to explain what he did and the officers would explain where he went wrong and what he did right. Farmers were often not very confident of their knowledge as often they would point out something they did and quickly apologise that they did not know if it had been the correct thing to do. I will give two examples of field days to illustrate. One of the examples is a field day held by a resettlement scheme farmer at Magazi while the second was a communal area field day at Madziva.

In both field days during question and answer time the experts asked more questions than the farmers. Two women, one of whom was a government worker, not a resettlement scheme farmer, each asked a question on two separate occasions. At the Magazi field day only two men asked a question and at the Madziva field day only three men asked questions and two of these asked more than one question. The experts asked more than 85% of the questions. Farmers did not appear to have any inclination to ask questions. Two farmers from Muringamombe pointed out that they were disappointed that AGRITEX only associated with the rich farmers at the expense of the poor. They would only see the AREX officers at large gatherings and field days where they sometimes could not ask questions and sometimes they needed the personal attention they had received from the AGRITEX in the early days. Except in just one instance, the people who asked questions in the question and answer segment were all good farmers by local standards.

The pattern of questions and answers on both field days confirms Bolding's (2004: 99) assertion that 'the actual performance of the field day often creates impression of a ritual, with each actor playing its assigned role of champion (Master Farmer), promoter of development (extension agent), benefactors (agribusiness company) or student (cheering audience)'. Experts sometimes asked questions they knew the answers to, such as asking farmers about the hectarage of a particular crop under cultivation and the fertilisers they had applied, highlighting what the experts thought was the relevant knowledge to be disseminated. At the same time, farmers sometimes also asked questions they knew the answers to, conforming to their role as the students who do not have knowledge. The farmer hosting the field day occupied an ambiguous position. He was the champion and at the same time, a student, there to learn from the 'experts'. Thus although fellow farmers sang him praise songs, he could not entirely conform to his role of champion as often he had to acknowledge his lack of knowledge and defer to the expertise of the agricultural knowledge 'expert'.[69]

It is clear that the stereotypes of the officers can determine the course of the field day. 'Professionals, like others, seek to order and make sense of their experiences. Like others, they construct realities, their interpretations and ways of construing the world' (Chambers, 1997: 33). At the first field day I attended, the officers thought that the people in the resettlement area knew how to farm and so did not participate actively. As discussed later in this chapter, because of the presidential elections that were to be held, the atmosphere in the resettlement area was highly politically charged. One of the officers pointed out that one had to tread carefully so as not to trample on other people's toes. Making enemies could be fatal. This difference in the politics of the resettlement areas and the communal areas was largely because of their different natures. As argued in an earlier chapter, resettlement was closely linked to politics. People were resettled by the ZANU (PF) led government after independence, whilst the people in the Communal Areas were not settled there by the government so they felt that they had an ancestral right to their land: as a result, they felt they had more secure tenure. Some farmers in the resettlement areas feared that, if the opposition political

[69] Freire (1993: 45), highlighting the lack of confidence by the peasants in their knowledge, wrote: 'Not infrequently peasants in educational projects begin to discuss a generative theme in a lively manner, then stop suddenly and say to the educator, 'Excuse us, we aught to keep quite and let you talk. You are the one who knows. We do not know anything'.

party won, the new government would expropriate their land and give it back to white farmers, forcing the farmers to go back to the impoverished communal areas. The communal area farmers had no such fears.

In the communal area, the AREX officer was highly active and indeed controlled the content of the field day. He did not hide the fact that he thought the farmers were ignorant and needed to be guided. Even the questions that the experts asked were different in the two field days. In the first they focused more on technical aspects like the amount of chemicals sprayed and spacing, while in the other field days they also had to ask about winter ploughing which they thought farmers were ignorant about. It was also left to the experts to point out what the farmers did wrong and warn against such practices.

Explanations by the experts were often put across in terms of local idioms and parables that farmers would easily understand and identify with. For example, when talking about farmers who reneged on loan payments the officers would start by talking of the dog that bites the hand that feeds it. The orator would involve the audience by asking what they thought, for example the person whose hand has been bitten by the dog he was attempting to feed should do. On one occasion one speaker likened the person who failed to service loans to someone who bites the back of a person carrying him to the hospital. As a result, he would be thrown down and left to die or to walk on his own to the hospital because he was ungrateful. Usually these parables would end with an appeal to the farmers not to bite Cottco on the back or on the hand.

Although much of the discussion on field days focused on loans and markets, crop pests and diseases were also discussed. These again were explained in terms that the farmers would understand. Sometimes farmers would take part in the unravelling of meanings so that they could understand better. For example, at one field day when the AREX officer was explaining a disease known as verticillium wilt one of the farmers in attendance just shouted that from the description the disease was just like AIDS. Although this comparison caused a lot of laughter, the officer then used the idiom of AIDS to explain the disease in the terms that the farmers could understand. If the stalk was pulled the disease would spread to wherever the soil from the stalk dropped. However although like AIDS the disease was incurable, it did not affect production. Put this way farmers could easily understand the explanations compared to if scientific jargon had been used.

When sitting and during the field tour men, women and the young did not mix. The men walked in front, the women following right behind and the young last. In Magazi the field day was held during the school term. As a result, the 'O' level agricultural class from the local secondary school attended. The students stayed close to the CCPS, who explained to them the various things related to cotton farming as well as the diseases that were likely to be encountered. Some students took down notes and some pretended to. During the question-and-answer session, the students asked nothing. On being asked why, they said they could ask their teachers at school to explain anything they did not understand. It was not proper to talk in the presence of elders and besides it would not bode well if a student asked a difficult question that the farmer was not able to answer.

During the field tour it emerged that both farmers and students explained or understood disease by referring to what they already knew. For example the leaves of the cotton crop both at Magazi and Madziva were turning red and purple in colour. On being asked why, both farmers and students were convinced that it was red spider mite. Although the farmers could not see the red spider mite on the leaves that was the only thing they knew that caused leaves to change colour. Some had even sprayed their cotton with red spider mite chemicals. As was pointed out later by the Cottco representative, leaves were changing colour due to a potash deficiency in the soil. This problem could be solved by using Compound K fertiliser that was high in potash.

Field days and social differentiation

In Mupfurudzi, as in most other Zimbabwean rural areas, it is difficult to talk of social classes but one can talk of social differentiation. When they first came to the resettlement areas most people had nothing and had to rely on the government for everything. However, some had a few cattle and some farming implements. Those who had a few resources managed to get a head start over those who had nothing, and in most cases, those who started better off are still maintaining the lead. This differentiation is not only between individual families, but also between sexes. It was reinforced at field days, although instances in which this differentiation was challenged also emanated from the way of the field days themselves. In Magazi, social differentiation was recognised in the way people were introduced as well as where they were eventually asked to sit during the proceedings. This is what Mr Gweshe, a local farmer and Cottco Representative at Magazi, said when opening the field day:

> Everyone should come here except those who are cooking (who as usual were all women). I also want to arrange people in the tent to sit according to their positions. Village heads, Gold Club members and AREX officers. Mr Nyamaharo come here in front. How can you sit in the back you who teach us how to farm? All gold club members come in front. You are the real farmers. We do not want you to get sunburn. The Chairman, Kahari and Kambiriyaenda come in front. All teachers sit over there (the other side of the tent which would soon be in the sun but it was still in front) and all health workers over here. All Branch and cell chairmen come here in the tent because we do not want you to complain that you were left out, as happened last time. I had not mentioned any of you including the District Committee members because I know that all these people are also leaders in Cottco. I know I am also one of them. (Clenching his fist and

raising it up high in readiness to make a slogan.) Forward with cotton farming, down with Tsenza[70] farming.

The question arises: is the field day a celebration of the farmer as I said earlier or is it just a celebration of the good farmer? All the farmers who belonged to the gold class were invited to sit in the tent where they would not get sun burnt whilst all the other farmers were not that important. What is surprising is that the wives of the gold class farmers (all male) sat in the sun and did not join their husbands in the tent. In the tent were only five women, I the researcher, my assistant, the cotton company secretary, the village health worker, and a Cottco group leader.

This assumption of the male head as the farmer and organiser and that women simply followed was challenged by the Mushamukadzi drama club that was composed entirely of elderly women. *Mushamukadzi* is a popular *Shona* saying that recognises that women are responsible for making good homes and their families, including making their husbands successful. Below is a summary of the main components of the Mushamukadzi drama. Throughout the performance all women were cheering, whilst some men were complaining that the women were an embarrassment as a drama club.

> The drama starts as the women attend a Cottco loan meeting. They are asked to bring their vouchers. Some have the vouchers some do not. Some of the women lack confidence and have constantly to consult their peers before answering questions from the loans officer. Some of the women are shaking and trembling. One family consisted of a lazy father Madzinga, and a hardworking mother, Nyekete. Nyekete is the one that went to get the loan. On the 22nd of October, she wants to start planting but is prohibited by her husband, who says the seed will get burnt. When the wife tells the husband that Mr Mushayi told her that that is the right time to plant cotton, the husband tells her that Mushayi knows nothing. When the wife is away, Madzinga sells the wife's fertiliser and chemical bottles because he says he wants to fix the wife. He uses the money to finance his and his girlfriend's drinking habits. Mushayi comes to inspect the field but the husband sends him away saying that the road is not suitable for a motor bike. He will first have to clear the way then call him back some other time. When the wife discovers that the chemicals are missing, she confronts the husband, who accuses her of not keeping the keys in a secure place, as the village is full of thieves. Nyekete accuses Madzinga of stealing the chemicals and she is beaten up. Madzinga tells his wife to sell to Tsikamutanda because he claims that Cottco grading system is unfair as they use a machine which when switched on turns all the cotton to a red colour. When Cottco comes to collect

---

70 The scientific name for Tsenza is Coleus Esculentus. In a paper discussing resource struggles in Kaerezi, Zimbabwe, Moore (1993: 395) notes that in Kaerezi, Tsenza was a women's crop but its cultivation was prohibited by the government because 'tsenza is widely believed to poison the soil, robbing it of its nutrients … (and) eroded deep slopes'.

the cotton, Madzinga tells Cottco that cows attacked the only bale they had. As usual the recovery clerks come and take the entire woman's property.

This drama highlights the problems that women faced in their farming endeavours. First they had no access to land and could only gain access to it through their husbands. If living with their husbands, the success of these women depended on the benevolence of the husband. The tensions between men and women were acted out albeit in a humorous manner. The intention of this kind of drama would be some kind of awareness campaign to condemn such practises. However, apart from the condemnation no solutions were offered nor does it seem that any were in sight.

The role of women at field days was particularly that of cooking and entertaining. Although women also took part in the field tours for some reason they did not seem particularly keen to ask questions but sometimes would congregate in groups, swapping stories and laughing. For example, after one performance at a field day in Madziva, women gathered to congratulate a group of people that had just performed. This happened although one of the officials was giving a speech. The women had to be asked to stop making noise and asked to come and listen to what the officials were saying.

Social differentiation was also recognised in the way the food was shared. The officials, as well as the gold class farmers and government workers, were given properly prepared food and drinks while the women and children got no drinks and some failed even to get food to eat. While it was conceivable that not all people were going to get enough to eat, it was also understood that the people in the tent could not go hungry. The stage was arranged in such a way that the performers faced the tent and had their backs to the women and children.

All farmers regarded field days as learning occasions. However, ten out of the fourteen respondents recognised the high entertainment value of field days as they mentioned that most people went to field days because of the food and beer as well as the entertainment although in the process they could learn a thing or two.

Field days, knowledge and politics

Although field days were primarily for disseminating knowledge, they were also highly political in nature and any person making a contribution had to realise this and in most cases use the proper political rhetoric. There were several different ways in which political rhetoric could be recognised. The context in which field days were held could also determine the politics and even the content of the field days.

At the Magazi field day, the political content was blatant. I argue that, although a field day was primarily a farming affair, failure to recognise the political hierarchies could lead to trouble. For example, the speaker, Mr

Gweshe, felt a need to introduce ZANU (PF) district members because he did not want a repeat of what had happened at another field day. At this field day someone had forgotten to introduce them and the District member had complained of having been left out and the speaker had to apologise. If one was a party official or a Sabhuku he/she was allowed to sit at the 'high table' in the tent regardless of whether the person was a lazy or a prominent farmer. Thus the field day was not only a celebration of the farmer and a disseminator of agricultural knowledge but also served to reconfirm the power structures in the village.

Village headmen were also party cell chairpersons and were therefore also introduced to the farmers although everyone knew who they were. They did not seem particularly keen to be introduced as party cell chairmen. This was probably because their position as Sabhukus gave them more than their role as cell chairpersons. As cell chairpersons their position had less power than those of all the other party members. This was because in the local and district ZANU (PF) hierarchies they occupied the lowest position and dealt only with their specific villages whilst the higher positions were usally district positions and much broader in scope. Cell chairpersons were also appointed by virtue of being a village head: they were not elected as were the other party positions in whicc people could claim power from popular support. As Sabhukus however, they had the backing of tradition and nothing could be done in their villages without them being consulted. There was some tension between ZANU (PF) district members and village headmen. During the 2002 elections and its aftermath, the power invested in the position of Sabhukus was eroded while party positions were consolidated. One of the Sabhukus was bitter that some party officials acted without consulting them as the representatives of the traditional authority. They overwhelmingly supported the party principles but felt they needed to be consulted more.

In Madziva, the ZANU (PF) councillor did not attend the field day. When his wife who had attended was asked why her husband had not come, she answered that it was because the Councillor had not been informed about the field day and as a result did not know about it. Clearly, the wife and all the people in the village knew about it, but the organisers of the field day had failed to recognise protocol and extend a personal invitation to the councillor as he felt his position warranted. Although the official function of field days is to spread agricultural knowledge by analysing the interaction of actors both present and not present at the field day one can thus gain an understanding of the political landscape.

Although songs and dramas were performed for education and entertainment the messages conveyed were also highly political in content. Even the speeches that were given also had a political inclination. The field day in Magazi intended to teach farmers about seed production. However, instead of sticking to the factual information on how seed was produced, they talked about the political and economic implications of land reform and why black people had to move into the field of seed production. Below is a speech by the AREX officer:

> Forward with Cottco. Forward with taking loans from Cottco. Our life is now leaning more towards the soil so we must not misuse the soil. We have enough land, now its time to show our farming prowess. Some farmers are always happy when they clear their loans, which is good but you are not yet a farmer if you only farm to pay back loans. One must have a lot of surplus. Crops like cotton bring foreign currency into the country. In the past we left this duty to white people but now that we have the land we must all be like Mr Kadungure (people clap hands). Farmers should plan their things on time and make the appropriate steps. A farmer should set targets for him or herself and follow everything that is required to be a successful farmer. Cotton has its own requirements. If you want to go into cotton blindly without following the various rules and regulations, cotton will bring you down. Some people planted at the same time as Mr Kadungure but did not follow all the requirements and their cotton is very embarrassing. Some people can even deny ownership of their fields if you ask them if the field is theirs, because the crops are a disgrace. A company like Cottco has created a good name for itself over the years. It helped a lot of white farmers to prosper. If you saw some of them riding around in Mercedes Benz, it was all Cottco's doing. Cottco gives enough inputs to people who are able to plan and use the inputs effectively. Some people cannot think. Some people are selling the inputs that they are getting on loan from Cottco. They deliver the inputs they sell at night so that they are not seen. At the end of the day, the kind of farming they do is very embarrassing. Inputs were sold here even during the days of AFC. People used to come here to buy seed and fertiliser cheaply. The only problem they would face was transport and at the end of the day we were the losers. (People started to cheer in agreement with the assertion.) We have a lot of companies here that are coming to buy our cotton, what have these companies done to help us? You know what children do when they see visitors. They start to behave in incredibly embarrassing ways. When the visitors go away the parents usually beat the misbehaving child. That is the same with farmers. In English they say biting the hand that feeds you. Let's say you have a dog and you want to feed it but then it bites you what happens?
> Audience: The dog will die of hunger because no one will want to feed such a dog.
> AREX Officer: You know what the white man did: he took the loan, farmed, and made huge profits that he banked. He always made sure that he serviced his loans so that he would be eligible for another loan the following year. Our problem is that we are failing to pay back our loans. That is dragging us as farmers down. We are encouraging you as your farming advisors to become very good farmers. When you arrive at the homestead of a farmer you can always tell. Farmers wear uniform. If you did not have a plan as you go from here today, you should start making plans. Forward with farming. What about those who do not know?
> Audience: They should be taught.

AREX Officer: No they should not argue (People start to laugh and clap hands whilst the AREX officer, Nyamaharo sits down).

There was talk that the Reeds (white farmers) were the ones that had dominated seed manufacturing that is why they drove Mercedes Benz and Pajeros. Although political rhetoric was a good appeal to emotions, it might not have been sufficient without also appealing to the economic side. The various speeches indicated that the white man was the yardstick, which the black man must emulate. Black people should cultivate seed like the white man, they should also be able to pay back loans like the white man, and they should also get very good harvests like the white man. The experts did not understand the logic of the farmer. Most maize farmers in the resettlement areas were farming to feed their families and then sell any surplus. Cotton farmers pointed out that they farmed cotton to raise money for school fees or to buy farming implements. As mentioned in Chapter 2, if a person could afford these, then he was a good farmer and did not have the desire or indeed the motivation to farm like a white man. For the resettled farmer, the yardstick was other farmers in the area, not some remote white man's farm. However, there also existed differences among farmers. The highly successful farmers, especially those who cultivated cotton and tobacco, aimed to farm more successfully than the white man and frequently mentioned that they could beat the white man if they had access to resources. Thus at the field days, appeal to politics and economy were all interwoven in the production of knowledge. Thus at the end of the day, knowledge was not only an accumulation of facts but knowledge was given a morality of its own and the advice given was regarded as something that a patriotic and principled person would adopt.

Context would also determine the kind of politics that was preached on field days. For example, an analysis of the content of songs that people sang at field days of the early 1980s and those of recent years reveal that the content is changing. Although the songs still espouse ZANU (PF) party ideologies, the focus has changed, possibly reflecting a change in the focus of the party. The first three songs are songs that had a political connotation sung in the field days of the early 1980s and they depended on the farmer's recollection. The last two songs were sung at the field days that I attended in 2003.

Our country Zimbabwe
That's where we live
We thank Muzhuzha and Mugabe
And all those who help us
To lift Zimbabwe.

Mugabe gave birth to a very small child
The name of the child is development

Development to the rural areas.

Is this what you planned for us
Planned for us our father Mugabe
We thank you for your plans
We are celebrating because of your plan.

Zimbabwe is independent
Our job is to farm cotton to develop our country
We thank Mr Jonga and Mr Mushayi
We thank Shava for progress.

Elders do not ever forget that we have taken the land
Give me embers
Give me dry grass
We will get the firewood at the big farms.

Although from the first to the third song farmers were singing about development, the cause of that development was credited to the president of the country and as a result he had to be thanked. Although the AGRITEX officer (who was Muzhuzha in the early 1980s) was also thanked, high level politics dominated the agenda of the field days. This could have been influenced by the context of the 1980s. In the 1980s the government gave people seed packs and fertilisers and helped them to start farming by farming a hectare for free for each resettled farmer during the first year. For instance there is a groundnut seed variety now popularly known as *KaMugabe* because people first cultivated the seed when they received it for free by the government. The farmer's achievements were attributed to the government, and since President Mugabe led this government, his name is frequently mentioned in songs.

However, in 2000 the focus had shifted. Instead of always mentioning the president in their songs, farmers are now singing about the people whom they frequently interact with in their day-to-day business. People like AGRITEX officers and agricultural company representatives are now being popularised in the songs. This does not reflect a wane in the popularity of the president but rather a decentralisation in the control over the day-to-day activities of the farmers. For instance, in the 1980s the government, through its various branches, was responsible for giving people loans of seed and fertiliser and sometimes gave them out for free, whereas nowadays people have to deal with commercial companies and banks to get loans. People also mention current political issues in their songs. In the last song, for example, people were singing about the fast track farms.

Listening to the field day songs one can understand some of the people's aspirations even though they are not explicitly referred to in the speeches of the different speakers. In the 1980s dramas performed at field days mostly focused on the problems farmers faced when transporting inputs from the producers as well as when selling their produce to the market. Dramas served as a way to convey grievances and problems to the relevant authorities, who were invited to the field days. These authorities would then be expected to come up with solutions. The reason for this might also be that most field days were organised by the AGRITEX department and there was no pressing need to advertise the products of any specific company.

The dramas performed in 2003, focused on gender conflicts as I mentioned earlier, advertised the company that was sponsoring the field day but above all they dramatised the various ways in which fake buyers could con farmers out of their crops. Below is an outline of a drama performed by a group of young people describing how people had been conned out of their crops the previous year by a fraudulent company that called itself Chorima.

Sineri Sineri Drama Club

(The Club consisted of 4 males and 3 females. The narrator started by saying:)

Ladies and gentlemen, we are going to present to you our drama entitled *anoudzwa odzoka* (Literal translation: Tell him/her when he/she is back, meaning, You sometimes tell people about things that will harm them after they have already been harmed, because when you tell them before they are harmed they refuse to listen to you). Some things are funny and some things make you sad. See for yourself.
Mr Mbanda: (shouting at top of voice) Hear me, hear me, people. Come to the meeting. Come to the meeting. Hey, the Kadungures come to the meeting (People laugh at the mention of Kadungure).
Mr Kadungure: Amai Tu, you have heard for yourself, let's go.

(At the meeting. The one who has been calling for the meeting asks Mr Kadungure or baba Tuji (Tuji's father) to open the meeting by praying. Every time they say baba Tuji people laugh because it sounds like babyish language for faeces.)

Mr Kadungure: Let us close our eyes and pray. God our Father hear us. God, we thank you because you created us as humans. If you had made us into cockerels instead of human beings, we would have to be killed for visitors on Christmas day. (Laughter and hand claps from the audience).
Mr Mbanda: Without wasting any of our time secretary, meet the people (Man 1 stands up). Women be silent: show that you like our programme here. (Amai Tuji and the other woman were whispering to each other.)
Secretary: Forward with farming. Forward with modern houses. Down with cars that do not have starters. (Laughter from the audience). I have travelled all over the world to many countries: therefore, I now know the countries to which we can sell our cotton at better prices. These other buyers are now denying you T-shirts and giving you shawls instead, but we will give you both (agreement from the audience within the drama). I will send our cars to collect the cotton.
Mr Mbanda: Make sure that the trucks are filled with cotton bales. In the interest of equal rights, I will leave you to madam. She has something to say.

Madam: Chorima hit and run (People start to laugh; apparently some people had fallen to the Chorima scam. This scam occurred when conmen and women came to the village and offered farmers prices that were unrealistically much higher than those of other official buyers. Because farmers were attracted to the higher prices, they sold their crops to these con-artists with the understanding that they would get paid as soon as the buyers delivered the crops to the actual buyers. The con-artists vanished with the farmers' crops and the farmers never got paid. The most notorious group of con-buyers was the group that called itself Chorima).
Mr Kadungure: It's harvesting money (*Kukohwa mari*).
Madam: That's not it. This is our slogan. If I say the *Chorima*, you say hit and run. We went to Italy. In Italy, they buy cotton at $10 000 per kg (equivalent to US$181 on the official market and US$17 on the illegal parallel market). For how much do these other people buy cotton from you?
Dramatists: (in unison) $50/kg (less than US$ per kg).
Madam: What is that? We give you much more. If you deliver cotton today, after three days, you get your weights and after one week, you get the cheques. You cannot wait for wealth to come to you. Women, let's get rich.
Mr Mbanda: You have all heard for yourselves. Mr Kadungure, can you say the closing prayer for the meeting?
Mr Kadungure: God, You heard what has been said here. Just keep on looking. Amen.

(Mr Mbanda and his wife argue where to take their cotton. In the end, they decide to take it to Chorima while Kadungure and his wife decide to go to Cottco. After some time, Mr Kadungure tells his wife that he heard rumours that all those who sent their cotton to Chorima were conned. He starts to laugh at Amai Mbanda, saying that when he went to their house to convince them not to sell to Chorima, Mrs Mbanda shook her chest at him until her breasts looked like they were going to fall down. She chased him away from her house (laughter from the audience). Amai Kadungure meets Amai Mbanda at the water pump and Amai Mbanda looks like she is wilting. She tells Amai Kadungure that they were swindled. Mrs Kadungure points out to her that it is because she never listens when others speak. Mr Kadungure is very happy and praises his wife for her prophetic abilities in insisting that they sell to Cottco. It comes as no surprise to him, as his wife's mother was a traditional healer. He ends up by saying he will never divorce her.)

(A new scene starts. The conman and women are enjoying themselves drinking beer and soft drinks.)

Secretary: That is why System (a local musician) sang that song where he says you were married to a crook. It is true you got married to a crook. We hit and run, baby. Let's go to the DRC, to Iraq, to Ethiopia (mentioning places that sound exotic to the villagers), to enjoy ourselves. We have got plenty of money to indulge ourselves. And just think of it. We sold the cotton to Cottco. (They laugh and move out of the arena.)
Mr Kadungure: (Goes to Mbanda's house) Amai Pinjisi (referring to Mbanda's wife), why do you look sad? Have you not yet received your money?

Mr Mbanda: The child has just been dismissed from school. We do not have money for school fees.
Mr Kadungure: Jah, you women of the *shava* (eland) totem speak too much (people are laughing: *shava* women are alleged to be verbally aggressive). I met those Chorima people in Bindura where they were selling your cotton to Cottco. I have been told to inform you that after ten days Cottco is going to send that fat man (people laugh because they know they are referring to Gweshe) to come and take all your plates.
Amai Mbanda: But they cannot do that. They already have some of our plates. They took when we failed to service the loan some years ago. I will tell them to keep those.
Mr Kadungure: They will take all your property because you have failed to pay them. That serves you right: you nearly ate me alive when I was trying to dissuade you from selling to Chorima. I feel pity for people like you. I will give your son his school fees and buy him a uniform. I will also pay back what you owe Cottco. Then I will take your two cattle. (They start singing and dancing signalling the end of the drama.)

Thus song and drama made sure that information was not only a one-way affair where only the experts disseminated knowledge. It meant that all those who were not given time to voice their opinions through speeches could still do so through drama and song, as there was no restriction on who could perform. For example, the drama highlighted that farmers were not happy with the cotton prices they got from Cottco. The dramas were also highly versatile as the various pieces could even include announcements that were made at the field day. For instance, Cottco had just announced that it was going to phase out T-shirts and hats for farmers and give them shawls and headscarves instead, a measure which did not go down well with the male farmers. This brings to mind Long's (2001: 7) assertion that creativity is not the monopoly of 'experts' and 'intellectuals', but that it is also manifest in the creative abilities of ordinary 'lay' persons and 'amateurs'.

Field days and social control

As was mentioned in the introduction to this chapter, apart from their education value, field days were also social occasions where people could gossip, catch on rumours, as well as settle disputes. Field days are also social occasions in the sense that what happens at field days is not only limited to the confines of the field day but often has a long history.

Disputes that arose at one of the field days I attended were related to issuing loans. Some villagers were dissatisfied with the way loans were issued and hoped that by airing their grievances in public they would get a speedy resolution from the dignitaries present. At a field day in Madziva one man complained that he had been denied a loan of cotton fertilisers and chemicals, because he had not cultivated cotton the previous year. Another man agreed with this, however, he said he had been denied the loan because he was told his name did not appear in the computer. This is how the dignitaries responded:

> Kahari: What happens is that the chairperson assesses your field. He has a rough idea of what to expect from your field. What do you do? You take two bales to Tsikamutanda, use two bales to service the Cottco loan, then hide the rest for other buyers.
> Mushayi: If you hide the cotton and sell it to other buyers, I will not give you anything. Why do you sell your cotton to people who did not give you any fertilisers and who sell you chemicals at exorbitant prices? If the chairman says I should not give you anything, then I will not. I get a crop progress report from the chairperson. If I count the balls in your field I will get a rough estimate of what you will get. Why should I get *masese* (second-rate beer) when others are drinking *musungwa* (best quality beer)?

Field days served as a form of social control to guard against any excessive abuse of powers by the chairmen, as discontented people could bring up their issues for discussion at the field day. However, it is not clear whether this was achieved to the satisfaction of the farmer, as in most cases the officers would end up blaming the farmer. In the second place, there were some social processes at play where the officers did not want to be blamed for any wrongdoing. Thus if there were any misunderstandings, it was either the farmer or maybe the chairman who was to blame, as they acted according to the information they received from the chairman.

On the other hand, when people attend field days they do not only do this in their capacity as farmers but also as villagers who have a life apart from farming. When people talk to you, they take into consideration your other identities as well. Thus when the young man complained that he had not received the loan fertiliser from Cottco although he had paid the joining fee, people chose to ignore his complaints, saying that he was not the one who had joined, but his wife. They said that he was not hard-working, that he did bad things to his wife, and wanted to control his wife's assets. Everyone knew that had it not been for his hard-working wife, he would not have anything to eat at his place. However, they could not bluntly tell him to get lost because they did not want him to lose face in front of his peers.

Disputes at field days did not only take place between farmers but could also arise between farmers and experts. Sometimes there was a latent tension between farmers and officers, and the farmers might feel they were being treated unfairly by the officers. For example, in one of the cases, the Cottco representative was complaining that some people had decided to skip him and go straight to Bindura to ask for inputs. This presented a challenge to his authority and he threatened to take action to bring the perpetrator to heel if such incidents continued to occur. This was what he said to the gathering:

> I am not very happy as I stand here. One of your chairmen here went to Bindura to ask for fertiliser. The person in Bindura is bigger than me but when I am within my jurisdiction, he is small. Do not think, because so and so married your daughter, that you can go to him and not follow the proper channel. I will reduce both you and the manager to your proper sizes. This year I am going to transfer that person to Bindura. He will get his inputs from there, as this is what he wants.

To be asked to collect inputs from Bindura was a bad thing, as transporting the inputs from Bindura was very expensive compared to collecting inputs from the local depot where one did not need to pay for transport. Furthermore, in Bindura, sometimes one would not get the inputs but was instead referred back to the field representatives. Thus although farmers could contest the authority of the officials, the officials were in a position of relative advantage and could use their powers to punish those who were recalcitrant. No one at the field day dared to say anything against what the officer was saying.

To better understand the position of farmers I had a conversation with one farmer on the issue of loans. I asked the farmer why he thought he was not going to secure a loan that year (from a different company than the one that had sponsored the field day):

> *You were talking about the loans saying you might not get the loans; why?*
> It's because Zengeza does not tell people the truth. He used to work for Cottco: now he is working for Agricom in Bindura. His job is to look for customers but he is 'very bad'. He told people that if they buy seed from Agricom, the organisation could loan them things. Now he told me that the loan facility has been closed already and I cannot have anything … but Taurayi told me today that they have not yet closed.

One village head also grumbled about some people who were making it difficult for them to get loans. Dhangeni, who resided in the village, had been denied a loan for no apparent reason. The headman maintained that they would leave whoever was doing this alone, but if he ever were to have a problem that would require the services of a village head, that he would punish him.

Women also used the field days as opportunities to discuss issues that affected them directly. These were not limited to farming but also related to the patriarchal structures that women always felt were very unfair to them. Mrs Gwaze, a very powerful woman in the party structures, upon being asked to say the vote of thanks alluded to the important role that women play in agriculture.

> Forward with farming. Cottco is the first wife. We now have the land that is needed: now it is for us to use that land. The wife is powerful. She is the farmer. These men are in our hands. If the wife is lazy then there is no development in

> that household. When the women and children go to the fields for an honest day's work the fathers and husbands remain at home sleeping. Some men are just men for what we see inside their pants. Such men do not know how to use their hands: they are just bulls (*people laugh*). We women are the ones who do all the work. If the child asks his father for a pencil the child is told to go ask her mother. We do not want the role of women to be recognised only in the singing and dancing at field days. It is we women who organise and attend the agricultural shows: men just want us to cook *sadza* for them so that they can go to the beer hall.

Women were also the ones who first started clamouring for women to stand beside their husbands when the husbands were explaining to the audience how they had managed to get good crops. The reasoning was that the women had contributed most of the labour, something that had to be recognised.

At field days, the role of women (as labourers and care givers) in agriculture is recognised and women seize the opportunity to act out the tensions between themselves and the men. However, this does not extend beyond the field day. For example, the woman who gave the vote of thanks chose not to confront the fact that some men do not work as much as their wives but in the end control the products of the labour. Instead, she chose to regard it as one of the strengths of women that they work to feed their families successfully, even in the face of useless husbands. At the same time, men accept that women generalise on their faults at field days, but would not take kindly to women who would try to contest such inequality – real or perceived – within the home. Overall, men in their capacity as household heads decided what was and was not beneficial education for the women. One old man pointed out that after attending lessons to which only women were invited, women usually told their husbands what they had learned. If the husband did not agree with his wife's newly acquired knowledge, he would just tell her to forget it. As long as women knew how to behave properly in their homes, men did not have a problem with anything women might say at public gatherings. As long as the education women received enabled them to live harmoniously with their husbands, men did mind women being educated.

As mentioned, field days were also occasions where people could catch up on gossip.[71] Women usually gathered in small groups, talking in whispers, laughing out loud, and sometimes clapping hands in an expression of

---

[71] In a study on gossip in Wedza Shambare and Bourdillon (2002:13) point out that many people claimed gossip to be useful. While people may spread gossip out of malice, or for sheer entertainment, gossip can be useful in spreading important information about things that cannot easily be brought into the open. Gossip criticises behaviour that is regarded as unacceptable within the community, and to this extent gossip helps to maintain the values of the community.

disbelief at what they had heard. On one occasion, a group of women who were cooking lunch were gossiping about certain women who they said never helped cooking when there were large gatherings in the village. These women had the notorious habit of coming to the gathering when it was already lunchtime so that they could eat the food cooked by other women. A nearby store had also been robbed and some people were discussing the event and pointing to certain young men within the village whom they thought were involved in the robbery. It was said that the police were about to pick the youths up so that they could assist them with their investigation into the robbery. However, invariably such stories ended with the storytellers cautioning the listeners not to tell people they had heard the story from them.

In one of the incidents, a group of people was talking about a young woman whose two children had died. People were discussing how unfortunate the young woman had been. Someone then said she had heard that the children were dying because the husband had not paid the motherhood cow. The discussion degenerated into how the young woman's husband's family was useless and how at one time the young husband had stolen a goat from one of the villagers. As it turned out later, the girl was paying for her refusal to listen to her parents, as her parents had never supported the girl's intention to marry into that family, because they suspected that the mother of the man now deceased was not a good woman (meaning that she was a witch). Field days are not only occasions where people gather to gain knowledge on farming; they go there to get information about other things as well.

Conclusion

This chapter shows that for a variety of reasons field days are important events in the calendar of the farmer. Agricultural knowledge is disseminated from the experts to the farmers and the farmers get an opportunity to interact with the experts in a relaxed atmosphere. Farmers can also seek clarification and solutions to problems they face. For the 'experts', field days have the advantage that they can reach a large audience at the same time. Despite being agriculturally inclined, field days have a highly political component that should be understood if the various actors within the context of the field day are to achieve their plans. Therefore, the politics of rural development have to be taken into account. Knowledge is not 'neutral' to be disseminated in a 'neutral' environment. Field days are not only organized for the dissemination of knowledge and entertainment, but are also occasions where status is reaffirmed and contested.

There is an inherent tension between AREX, other 'experts' and farmers. In Chapter 5, AREX was blamed for the failure of Katsoko seed and blue fertiliser, as they were accused of siding with the Large Scale Commercial Farmers (LSCF) and agro-business against the government. In this chapter,

AREX was said to favour the rich farmers and accused of not offering the personalised services that farmers wanted. This tension was a result of the different worlds in which farmers and AREX operated, giving rise to diverging worldviews. This tension was exacerbated due to a lack of resources from the side of AREX to implement government programmes. Where there was a discrepancy between government policy and what AREX did, farmers were quick to blame AREX, whilst AREX pointed to its lack of resources.

Because of their rigid application of scientism, AREX and other knowledge experts fail to understand the social context within which farmers operate. 'All forms of external intervention necessarily enter the existing lifeworlds of the individual and social groups affected and in this way are mediated and transformed by these same actors and local structures' (Van der Ploeg and Long 1994: 64). AREX and other experts fail to recognise that social relations within any community can affect the status ascribed to knowledge. Whereas farmers consider a farmer within a social context, AREX only considers productivity, which can lead to the selection of unpopular farmers to host field days. Farmers did not agree with AREX' selection method of good farmers, which in turn had a negative impact on knowledge dissemination. Knowledge should be understood within a social context if it is to be disseminated successfully to the intended audience.

For farmers, field days were not merely occasions for gaining new farming knowledge. People attended field days so that they would not be regarded as lazy, to maintain good relations with other villagers, for entertainment, to settle disputes and to a lesser extent to gain knowledge and information.

Farmers did not ask many questions and most of the discussions were directed by AREX and field day sponsors. This might have been because farmers were not participating in their own projects but in the projects of the experts. Field days were mostly initiated by AREX and agri-business. These field day meetings were different from village meetings initiated by the villagers, where the villagers directed their own discussions and fully participated (see the discussion of such a meeting in Chapter 5).

The gendered nature of knowledge made itself apparent on field days. The men were the Master Farmers, whereas women could only be their wives, mothers or daughters and never Master Farmers themselves. This was the interface between culture and knowledge. In spite of their knowledge, women could not be regarded as Master Farmers in the presence of their male 'guardians', and were thus relegated to a perpetual state of ignorance. The different forms of prevailing social relationships mediated the status of knowledge.

Observation highlights that an ability to tread the delicate ground between politics and economics was essential for the dissemination of agricultural knowledge at field days. Thus in this vein it can be concluded that knowledge is not a resource and an artefact out there waiting to be used, but a social construct and a social relationship. 'Knowledge emerges only through invention and reinvention, through the restless, impatient, continuing, hopeful inquiry human beings pursue in the world, with the world and with each other' (Freire 1993: 53).

7

Knowledge and practice: Men, women and children

Photo 7.1 Mupandasekwa's nephew's wife shelling groundnuts with her two daughters. Photo by Christine Kwangwari (research assistant).

Introduction

Households should not be considered as single units but rather as containing people with different needs. These needs can sometimes converge depending on the context and situation that household members find themselves in. In a study of a sugar out-grower's estate in Zimbabwe, Mate (2001: 44) noted

Photo 7.2 Father and son helping each other to smear paraffin on chickens to get rid of ticks. Photo by author.

that 'the effects of lack of information and knowledge were revealed in those households where the original settlers are deceased and the widows had to be in charge'. After the death of the male head these widows had to rely on male neighbours for information and knowledge on how to do certain things.

Thus any study of knowledge can never claim to be complete if the different groups in what has been traditionally known as the household unit are not taken into consideration.

Although differences may exist between the different groups within households, for analytical purposes I have placed people within the households into broader groups depending on gender and age. Knowledge can only be fully understood when men, women and children within households are considered separately, in order to bring out what they know and consider as knowledge. Men, women and children may have different knowledge, not only because they are situated differently with regard to their access to knowledge, but also because their social positioning and the parameters within which they operate can differ. On the other hand, the knowledge that people have or think they have, impacts on how they process information and knowledge they receive from elsewhere.

Households are characterised by differential access to resources for men and women and for the young and the old. Within households,

> ... relations between people are transactional and punctuated with negotiation, bargaining or open conflict. Power relations determine resource flows, whether they emanate from socio-cultural values or material endowment (Mate 2001: 58).

The household is not an altruistic entity. According to Shepherd (1998: 15), the concept of the altruistic household needs to be challenged because it allows for 'development theories to be targeted unthinkingly at the (usually male) head of household; the notion of shared poverty when in fact poverty may be experienced differently by men and women'. As discussed in previous chapters and later in this chapter, even the dissemination of knowledge as part of the development initiative was structured in such a way that women were frequently disadvantaged in accessing formal information. In the same vein as Shepherd (1998), who advocates studying the different members within the 'household', I evaluate the different groups within households separately in order to understand their approach to knowledge and practice.

Reconstructing the household in this way provides an understanding of the needs of the household unit as apart from those of the individual members, showing how such differentiation can have an impact on what the different members know or think they need to know. The issue of gender and knowledge is a critical one that requires careful study. In a study of newly introduced rice technologies aimed at improving rice cultivation in Gambia, Carney (1993: 337) states that

> ... development meant the delivering of female labour for intensified rice farming without concomitant income gains. The reinterpretation of customary tenure by male household heads and village elites aimed to ensure continued female access to rice land, but only as workers on plots whose benefits would flow to men as disposable surplus. The donors' uninformed view of the Gambian household-based production system was to prove the nemesis of the project.

How women perceive new knowledge and technologies brought in and adapted by their husbands or other male heads can be very significant in whether that technology fails or succeeds. Therefore, how certain technologies and knowledge affect, and are affected by, household dynamics, especially those based on gender, should be taken into consideration when trying to introduce 'knowledge' from outside.

Men and women sometimes operate under different constraints so that they may reach different conclusions or take different courses of action. That is, the different constraints that the genders face may have an impact on what they know and what they do. As mentioned in earlier chapters, the land resettlement scheme was conceived of in terms of male household heads and female labourers or housewives (see Jacobs 1993). All of the female household heads were widows. The general picture of the female-headed households is grim, if we take into account that at the national level, according to the Poverty Assessment Survey (1995: 69), the majority of the female-headed households were placed in the very poor category: 57% as compared to 40% of male-headed households. Furthermore, only 28% of female-headed against 40% of male-headed households belonged to the better-off categories. Compared to the 40% of male headed-households that were not in the poor category, only 28% of female-headed households belonged to this category. Elsewhere, Mombeshora (1998) makes the interesting observation that the poor are the most unwilling to take risks and adopt new technologies.

Knowledge dissemination and formal channels

Agricultural lessons

AGRITEX officers formally impart knowledge about agriculture, through periodic courses of formal lessons for Master Farmer certificates. As mentioned earlier, being a certified Master Farmer can mean the difference between being allowed to access some resources from AREX and being denied this, because some government resources are channelled through the Agricultural Research and Extension Department. The government adopted this strategy to avoid a high number of bad debtors by giving enough fertiliser and seed loans to those farmers who were vouched for by AREX officers. Therefore, as discussed in Chapter 1, some people attended the formal lessons so that they could establish a good relationship with the agricultural officials. This good relationship would ensure that the AREX officers would put in a good word for them to obtain the loans. None of the female heads in Mupfurudzi had ever attended the lessons offered by AGRITEX officers while only one of the wives of the male heads had attended the Master Farmer training lessons. On the other hand, all male heads claimed to have attended Master Farmer Training lessons at some time.

The high illiteracy rates among women were usually mentioned as one of the reasons why most women chose not to attend the lessons. Two female household heads in Muringamombe had this to say when asked why they had never attended the formal lessons:

> It was like a school. You have to read and take notes and remember everything. I would not have been able to cope. It was a school, where those who passed were given certificates and resettled *kuminda mirefu* (long fields).

The second woman said:

> I am not able to write. In addition, the classes are always full of men and it's a little embarrassing to be the only woman in a class full of men.

However, further investigations indicated that although illiteracy could have contributed to the women's lack of confidence to attend agricultural lessons it was not the only reason, as many illiterate men had attended the Master Farmer Training lessons and managed to acquire the certificates. The AREX officer, Mr Nyamaharo, claimed that literacy was not a valid reason for not attending.

> You know that in Wellaway we have Bosiya Tiriboyi. He cannot read and write. He was not a very good farmer and people used to laugh at him. After attending our training programmes, he is now a very good farmer so that very few people beat him. When people began to see that he was prospering because of his association with AGRITEX, they also began to attend training sessions.

It should be noted that the agricultural officials also taught oral classes and gave oral exams for those who could not read and write. Consequently, the failure of women to attend the lessons may be related more to the gender division of space in which the public domain is reserved for males and women would be embarrassed to compete for that space with men. Cheater and Gaidzanwa (1996: 191) point out that in Shona societies 'traditions of male mobility contrast sharply with female immobility'. Women who were highly mobile and ventured out of the bounds of their immediate residential neighbourhoods were often labelled as prostitutes. Such negative associations between female mobility and prostitution might have militated against women attending agricultural lessons that were often held far away from the village and sometimes entailed being away from home for a few days.

In some cases where females attended the lessons, they would meet with a variety of obstacles to obtaining the Master Farmer certificates. As mentioned in Chapter 1, in the past, after going through the necessary training a person had to build an implements shed before being awarded the Master Farmer Certificate. It may have been requirements such as these that discouraged some women to go to the Master Farmer Training Programme, as they would need approval from their husbands to initiate certain developments on their homesteads and this approval was not always readily given. One woman who had attended the Master Farmer Training for a number of years but had still not managed to get the certificate explained why she had failed to implement some of the things she had learned:

> What happens is that if we who wear dresses go to the workshops, if we come back and try to implement what we have learned, there will be conflict in the household. When I came back, I tried to have the shed built, but my husband insisted that he had never seen that since he was born. He kept asking why a plough should be kept in a built shed. He could not understand it. There was a lot of conflict until I decided to drop the shed issue. As a result one can not implement everything one learns.

Thus illiteracy as the main variable explaining the non-attendance of women is evidence of a short-sighted view and fails to take account of the involved household dynamics.

It was not only the agricultural lessons that women did not attend, but also other agricultural meetings that were held in the village. Although these were usually held on Fridays (*Chisi*), women could still not attend, as they used this day to focus on the domestic tasks they had ignored during the week. It was usually on Fridays that women cleaned their houses, redecorated them, washed the family's clothes, and watered their gardens, whilst they also had to prepare meals for their families. On the other hand, men would be visiting friends and those who liked to drink would spend the Fridays at the local bar talking to friends and generally lazing around. It turned out that men had more time than women to attend these meetings. The AREX officer confirmed the high attendance rate of men as compared to the low attendance rate of women for both agricultural lessons and agriculturally related meetings. He blamed this on the fact that women are usually busy with domestic chores and people are also used to the fact that only men attend meetings and make the decisions. However, because they never attended any meetings women could not get enough information to understand the entire production process instead of just the provisioning of labour.

> Yesterday I was in DERUDE carrying out a crop census on behalf of the central statistical office where I have to ask farmers how much they produced this year. As it happened many male heads died. So I had to ask the women and their answer was invariably that 'we do not know anything: father is the one who knows but unfortunately, he is dead. We only know how to work'. At the end of the day, we just guess. That affects the accuracy of our statistics.

Because of the gendered nature of access to public information, women lacked competent knowledge in certain aspects along the agricultural production chain.

Youths and knowledge

Young people are an important element in the dissemination of knowledge and the adoption of technology. We talked to fourteen families on the issue of youth. All families thought children were important for information gathering. Seven of the fourteen families had sent their adult sons to attend agriculturally related meetings on their behalf. One of the young men was now overseeing the running of the family farm, although still in consultation with his father.

All the female household heads in the sample, except one, had adult sons whom they usually relied upon to attend the meetings and lessons and to pass on the knowledge to them. However, it was not only the female households that relied on their male sons but also those households where the male head was very old. Adult sons are very important for many households in order to obtain outside knowledge as their mobility is high as compared to that of old men and women. In only one household an adult son mentioned that he did not attend the Farmer Training Lessons, his reason being that they were taught in Shona. He maintained that he would have attended if the lessons were conducted in English as they were at school.

Youths were also important in the dissemination of knowledge and technology, especially adult sons who had no limitations placed on their mobility.

> *Do you think you can learn anything important agriculturally from your children?*
> A child can tell you something that he has seen somewhere. Sometimes he can encourage you to do it; sometimes the things work out well sometimes they do not.
>
> *Have your children ever taught you anything that you did not know?*
> My children have never taught me anything. I am the one who teaches them things because I have always had the knowledge from the farms where I used to work. I was a tractor driver and I used to live with white people.

The importance of mobility for knowledge acquisition can hardly be over-emphasised. For example, some youths maintained that they had heard of Agricom first when they had gone to Bindura town to visit friends. They had gone to the offices to inquire about how people could secure loans from Agricom. When they came back to the village they told others, who immediately teamed up to secure loans from Agricom and made one of them their group leader because he had the most knowledge on how Agricom operated. Although male adults might deny having learned anything from young people, it is undeniable that they are important for external information gathering.

Sometimes, like in the above quotation, adults contradicted themselves, pointing out that they learned something new from their children and then denying it later. This might be because of the cultural considerations in which older people are regarded as wise and are expected to impart their knowledge to the young instead of the other way around. Thus when asked a hypothetical question it was easy for old people to admit that they did learn from the young but when a direct question on the same issue was posed the answer was almost invariably a firm negative. On the other hand, women had no problems in admitting that they were learning from their children since they had no cultural constraints forcing them to appear more knowledgeable than they actually were.

The importance of having highly mobile male children unencumbered by societal obligations as compared to their female counterparts, indicates how the development of human capital, which is a key element in development, was skewed in favour of men. Although many researchers (see Rukuni 1994: 19) correctly link the development of agriculture to technical research, extension, price support, marketing infrastructure and finance, this concerns the national level where the production statistics focus on what is produced but do not differentiate between who produces what in terms of gender. At the local level there is an invisible thread linking knowledge to development at the household level.

The youths were not only important as information gatherers but also in the implementation of knowledge. Knowing is not an end in itself; knowledge should be brought into practice to produce results. Some people in the village who were known to be very good farmers did not appear to do so well after their children had moved away to get jobs or to start their own families.

Although the importance of youth is not immediately apparent, their importance in household food security can never be overemphasised. Some households that did not have grown-up children suffered food shortages, as there were usually not enough people to provide labour and bring in new ideas. Of the seven poor households, four household heads mentioned labour constraints as also contributing to their low crop production. One of these four household heads had no children, two had young children who could not contribute meaningfully to labour and the forth household head had six adult children but they were not contributing to agricultural labour, as they were all working elsewhere. These children also did not remit any of their income to their parents. One of the medium-wealth households whose general wealth levels were declining had lost some of its adult members to death and was left with young orphaned children who were too young to provide substantial labour.

It is not only the young adults who are important to the household as far as knowledge and food security are concerned. Although the very young do not contribute new knowledge, they acquire agricultural knowledge through performing small tasks for their families. The young often run small errands: they may lead the cattle during ploughing and also herd cattle. In one household where all the adult children were temporarily unavailable, a young eleven-year-old boy skipped a day of school so that he could have the cattle vaccinated against black-leg. On other occasions he also had to miss school so that he could take the family herd to be sprayed for ticks. Thus the young also learned from their parents how things were done as they grew up.

There were sometimes conflicts between the young and old, since the old, who claimed they had more knowledge because of their seniority, sometimes wanted things to be done in certain ways whereas the young, who also claimed knowledge by virtue of having been taught at school, wanted to do things differently.

Apart from gaining access to information through their adult sons, women could informally acquire information from friends whose husbands had attended agricultural lessons and meetings. Two female household heads mentioned hearing about new information in this way. Women would also sometimes discuss what they had heard with other women at the water pump.

School lessons

Access to knowledge for the young was divided along gender lines, like it was among adults. While most young men regarded schools as an important source of agricultural information, young women depended on the radio. Girls had a high drop-out rate from school because of pregnancy or a lack of money. Usually when there was a shortage of school fees girls were more likely to be pulled out so as to enable the boys to attend. Until recently the law was such that in the event of pregnancy a girl would be expelled from school as her pregnancy was thought to disturb and agitate other students. Although the law has been changed, parents are still likely to pull their child out of school if she gets pregnant, as a way of punishing her and sometimes just so that she can look after her baby. During the course of the study only one girl from a sample household fell pregnant. Her parents pulled her out of school although they said that the girl would go back to school as soon as her child was old enough.

Although at school students are free to take up any subject they want regardless of gender, females gravitate more towards traditionally defined female fields whereas boys are inclined towards the traditionally defined male fields. As mentioned in Chapter 4, some agricultural areas and tasks were gendered. For instance, large livestock were regarded as a male concern, and women were afraid to experiment with these lest their experiments failed and they would be blamed for any losses incurred. Thus girls usually took up fashion and fabrics while boys did agriculture, building and metal work. Because women were not resettled in their own right (Gaidzanwa 1995; Jacobs 1990 and 1991; Rukuni 1994; Mate 2001), it made sense for women to take up sowing because although they could have done agriculture at school, their decision-making in agriculture was limited as they could only have secondary access to land. On the other hand, if they owned a sowing machine, they would be in full control of their machine and the products made with its help, as sowing was a female field and men would rarely challenge them.

The young were more comfortable with knowledge they got from school; people could for example use their school knowledge on plant spacing, fertiliser application and crop varieties. However, all the youths pointed out that sometimes school knowledge was too scientific to be of practical use, for instance, the thumbs rule to ensure that adequate rain had fallen. Sometimes they also modified fertiliser application from what they had learned at school to suit their own conditions. Thus although the youth held school knowledge in high regard, it had to be modified like any other knowledge to suit local conditions and available resources. School knowledge however, was regarded the same as knowledge from AREX officers.

Although all male heads denied learning anything useful from their children, everyone pointed out that school knowledge was important and very useful. The adults maintained that they learned nothing from their children, not because they thought that their knowledge was useless but because they felt they had already learned everything a long time ago at the commercial farms. All female heads, however, were very willing to learn new things from their children. They pointed out that people with children that had received agriculatural training at school were doing well and even the successful farmers in the village had received agricultural training of some sort. For women, school knowledge is reliable and can be depended upon. Thus it can be witnessed that youths in female-headed households have more freedom than youths in male-headed households to implement their ideas and knowledge.

At primary school, both boys and girls did some agriculture. For instance Muringamombe Primary School had a school garden and a school field, where pupils were expected to work during school days as well as holidays. Mr Togara, one of the teachers at Muringamombe, was very excited about the garden and field.

> We are training the Master Farmers of tomorrow. Some of these children will not be able to go to secondary school. So if we can teach them to farm, we will have given them a skill they can use to look after themselves in the future.

Parents did not take this primary school agriculture seriously since they regarded it as a school fund-raising venture. All the children did was to provide the physical labour. Some teachers also had the habit of instructing the children to work on their personal fields, shell maize or groundnuts, and sometimes prepare peanut butter for them. This was done under the guise of teaching practical skills to the children. The parents regarded it as mere exploitation, although they never opposed it. This was so because the children enjoyed what they did, for if they worked for the teacher they could get lunch. If the child was a good worker, good relations could be established between the parents and the teacher, so that if the parents needed a cash loan they could easily approach the teacher.

Consulting children was also based on the age of the child, with those in primary school rarely being involved in decision-making. Young children in primary school were generally regarded as not having knowledge.

Gender

After marrying, a woman's knowledge becomes redundant as she has to adopt the ways of her husband's people. Sometimes during the interviews with both husband and wife, the husbands expressed their surprise at things their wives claimed to know.

> *(Christine) What of sunflower? Why do you cultivate it?*
> Mr Mbanda: We grow sunflowers for sale. We also use sunflowers for making cooking oil.
>
> *Are there people who have these oil-making machines in this village?*
> Mr Mbanda: Yes there is someone with the machine at Danken.
>
> *Danken, is it a village or a farm?*
> Mr Mbanda: Danken is a village just like this one. That is where the person with the machine resides. If you have sunflowers or groundnuts that you want to be processed into oil you just pay them and they will do it for you.
> Mrs Mbanda: But I can make some oil.
> Mr Mbanda: Iii ... (In a voice mixed with surprise and disbelief) you can make oil!
> Mrs Mbanda: Yes I can: my mother taught me how.
>
> *So how do you do it?*
> Mrs Mbanda: You just pound the sunflower in a mortar. After that you process the cooking oil from the sunflower just like you do groundnuts. My mother used to do it and she taught me.
> Mr Mbanda: This one knows since her mother taught her (still not convinced).
> Mrs Mbanda: My mother used to do it, but since I came I do not do it because the sunflower is mostly grown for sale.

Women's knowledge of agriculture may remain unutilised when they move into a new household, as in the case of Mrs Mbanda, despite the fact that she and her husband had been married for many years.

The redundancy of women's knowledge was not limited to the older generation. One married young woman in her late twenties admitted that after she got married she had never used any of the agricultural knowledge she had acquired from school. Instead she just followed her husband's way of doing things.

> … as you know, if you finish school and get married soon afterwards, you start to think all the things you learned in school are not important. You do not even want to think about whatever it is you learned, and if you do, in most cases you do not even want to use it in case it conflicts with what your husband wants. The husband is the government and one does what government wants.

Women preferred not to practise what they knew because they did not want to be blamed in case their knowledge failed to work and jeopardised the family's food. As discussed in Chapter 4, the wives of an absent male household head did not use a method that would have saved their ailing ox from death, because they were afraid to do that since they would be blamed if the cure failed and the ox died.

Although all household heads concurred that school knowledge was important, most male household heads denied that they learned much from their children. All female heads in the sample claimed that they had successfully utilised most of what their children had learned from school and they had found it very helpful.

When it came to farming, men and women often knew different things and therefore often clashed when making decisions. For instance, because they sometimes utilised differing knowledge sources, men and women in the same household from time to time reached incompatible conclusions on how the family farm should be cultivated. For instance, in one case the wives of one man wanted to inter-crop while the man said that this was not a good farming practice like he had learned long ago at the big farms and the agricultural lessons. In the end the wives mixed maize with cow-peas and pumpkins and the husband ploughed the cow-peas and pumpkins down. He was furious that his wives had risked the family food basket by refusing to listen to his good sense.

Making decisions

Processes of making decisions help us to understand the gendered nature of knowledge. In nine out of fourteen cases in the sample, household heads claimed to consult other members on the selection of maize variety. This information was not always reliable: in one case the head said he made decisions alone and then when his wife was present said he consulted her; and in another, the husband claimed to consult his wife while she denied this when she was alone with the interviewer.[72] Three of the four female heads consulted their adult sons or other male relatives before deciding which seed variety to plant, especially where maize was concerned. In the fourth female-headed household there was no real consultation as the grandson who worked in Harare just bought whatever seed variety and fertiliser he managed to secure and gave them to her. Although sometimes the grandmother would have preferred some other variety, she was still thankful for the seed that she got and she never complained to her grandson. On the other hand five male heads maintained that they consulted their families.

Women in all households were involved in decisions affecting the choices of peripheral crops such as groundnuts, roundnuts, *rapoko* and Open-Pollinated Varieties of maize, which are regarded as women's crops. This was a recurrent theme throughout interviews with the different house-

---

[72] Goebel's (1999) survey in the Wedza area of Zimbabwe has 70% of men deciding on the planting of maize.

holds, indicating that knowledge has gender. It can be argued that knowledge was gendered to the extent that female expertise was rarely questioned in areas that were traditionally considered their domain. However, when it came to farming crops like maize, cotton and tobacco men generally regarded themselves as more knowledgeable. Men's advice was also more actively sought with regard to these crops. In all interviews, both women and men pointed out that crops like cow-peas, round nuts and ground nuts were women's crops and women had more knowledge where those crops were concerned. What is interesting however, is that women also claimed expertise on the male crops whereas males always deferred questions on female crops to their wives, professing complete ignorance where these crops were concerned.

On occasions when women were consulted on cash crops this was usually not done because their knowledge on these crops was highly valued. All women in the sample concurred that their husbands would ask them for their opinions but when their opinions did not agree with those of their husbands they ended up doing what the husband wanted. When he started to cultivate flue-cured tobacco instead of the air-cured varieties, one farmer did not consult his wife because he thought she had no knowledge of tobacco farming. However, at first the wife resisted this crop because it was very labour-intensive, and she withdrew her labour by pretending to be sick for part of the farming season.[73] Although the husband strongly suspected her of feigning illness, he could not force her to work without seeming like a villain in the village. She later agreed to cultivate flue-cured tobacco because it paid well, but the husband now consults his wife before adopting any new technology because he does not want her to withdraw her labour.

Women who were said to have knowledge about crops like tobacco and cotton were usually widowed women. Especially concerning tobacco, people agreed that these women were usually free to attend training courses held by AREX and their success could not be easily be attributed to any male presence. Especially in one case a woman had successfully adopted tobacco farming after her husband's death. People claimed that it was because she had attended tobacco-training courses and gained the requisite knowledge. However, if the widow had adult sons any success she might achieve as a

---

[73] Ehrenreich (1973: 43) noted that upper class women in England during the industrial revolution feigned illness to avoid intercourse with their husbands. Women sometimes subverted the sick role to their advantage as a form of birth control; for women who wanted to avoid pregnancy, 'feeling sick was a way out'. 'A doctor could help a woman by supporting her claims to be too sick for sex: he could recommend abstinence. So who knows how many of this period's drooping consumptives and listless invalids were actually women, feigning illness to escape intercourse and pregnancy?'

cash crop farmer or indeed in farming in general, was quickly explained in terms of her adult sons.

To prove that women had less knowledge than men, one farmer pointed out that it could be seen by the yield that women got when they farmed alone. Even in the traditionally female crops like groundnuts, women still got less than men when men decided to venture into groundnut farming, because men could access knowledge from AREX and employ it in their farming ventures. However, one woman maintained that this was not because women knew less than men but simply that since men allocated the farming land they naturally allocated the most fertile land for male crops. The man who blamed the low productivity of women on their perceived lack of knowledge conveniently did not mention, as he had earlier mentioned for farming in general, that sometimes a lack of access to resources like money to purchase inputs might also explain low productivity among women.

However, although women feel they can do better if they are given good soil, they are generally not bitter about the way land is designated. This is because women's crops are not grown for commercial reasons, and although these crops allow for diversity in people's diets, they cannot afford people food security in the way maize can.

Women were in favour of Open-Pollinated Varieties whilst men expressed preference for the certified seed. All questions relating to Open-Pollinated Varieties were usually deferred to the wives of household heads during interviews. It was also the women who named Open-Pollinated Varieties because they were the ones that dealt most with these varieties and could name them after their characteristics.[74] The issue of Open-Pollinated Varieties was gendered because they were usually not cultivated for commercial reasons; hence they fell under the control of women.

Investments

The investment patterns of men and women differed to some extent. Out of eleven people in Mupfurudzi, eight said that while maize was an important crop they did not use proceeds from maize to buy any large items. Three of these eight households were female-headed and they claimed that they could not use money from maize for any large purchases since they only rarely got enough to feed themselves and their families. Most households, even male-headed ones, used money from maize to buy food and clothes, and to pay for their children's school fees, whilst some went further to use maize as pay-

---

[74] For example one open-pollinated variety was named *Kadya* (the small eater) because although the maize cob and even the kernel was of a small size, and multi-coloured if planted in the same field with other maize varieties, cross-pollination would occur and the characteristic of *Kadya* would dominate. Hence the metaphor of eating simply denotes that this variety would dominate all other varieties.

ment for labour they hired to weed their fields. Three people admitted to having used maize to buy things of great value. Two male heads had bought solar panels and the third person (a widow who did not grow cotton and tobacco) built a kitchen and a granary using maize as payment. These three also used maize as a source of food and cash for school fees and clothes.

Farmers were concerned that they were using most of their maize to pay back loans on inputs, and that they were left with no maize to invest in other things. Thus, whereas in the past people could invest cash from maize into their children's education, people are now looking more and more towards other cash crops like tobacco and cotton. As a result, female-headed households with no adult sons are getting poorer and have less and less to invest in their children's education and in the acquisition of implements. This is because these women have little access to knowledge that would enable them to diversify into other crops like tobacco and cotton. They do not only have less access to the information they might need to diversify into these new crops, but even less access to the requisite resources.

Poverty and the poverty of knowledge

Female-headed households predominated among the poor households in the village. Out of the four female-headed households in the sample, two were in the poor category, one in the very poor category while the fourth was in the medium-wealth category. The very poor woman had no children and had thus no access to labour, while the two poor households had all their children at home and had no other sources of income. The woman who was in the medium-wealth category had sons and grandsons who were well-educated and worked in the urban centres.

These women explained their poverty in different ways. One of the poor households maintained that they were poor because the male household head had been ill before he finally passed away and could not do much field work. The second household said that it was because the husband had many wives and many children and everything he worked for went towards family consumption. The other poor household attributed its poverty to the lack of labour. One of the good male farmers pointed out that the poor households were very poor because they did not have the required knowledge to farm profitably. A similarity among these households was their lack of access to resources when they came into the resettlement scheme. They owned no cattle and only one household owned a plough. As mentioned earlier, most of the medium-wealth households came into the village with a few resources of their own that gave them a head-start over other households.

Women might also have predominated among the poor households due tof their lack of mobility. In the sample only one female household head had regularly attended meetings. None of these female household heads had ever gone to the GMB depot or to Cottco in Bindura where some respondents claimed to get information on new developments in farming by talking to farmers from other areas or to the employees of these organisations on an informal basis. A highly mobile woman headed one of the progressive female-headed households (not in the sample). This household head was a traditional healer and almost everyone in both Mudzinge and Muringa-mombe claimed some sort of relationship with this woman. She attended most of the agricultural meetings, which other women shied away from, and in 2001 she diversified into tobacco cultivation, a crop which she had not cultivated when her husband was alive. Although other factors might explain her success one factor that stands out is that compared to other women she was very mobile and could obtain information that other women could not. In the same village there was another female traditional healer (in the sample) who did not have a husband to limit her movements, but who was not a successful farmer and barely managed to have enough to eat. Although she was confident and as mentioned earlier most people in the village affectionately referred to her as *ambuya* (grandmother or aunt) she did not actively seek information on how to access resources and loans from outside and did not attend lessons to learn about new crops. Mobility was a critical factor, especially in households that did not have grown male children to take over this function of gathering information from other sources.

Conclusion

Knowledge dissemination is gendered, as women and men often use different sources of information. Few women attend Master Farmer Training programmes and although a large number attend field days (see Chapter 6), it is mostly as entertainers. Space is divided into specific gender domains as women sometimes do not feel comfortable attending male-dominated meetings. Illiteracy and social powerlessness among women also determine what women know relative to men. Thus by understanding household dynamics, the weaknesses of the official channels of knowledge dissemination can begin to be appreciated.

The gender division of labour also implies that men can frequent places and attend meetings that women cannot because they have domestic chores to attend to. The division of labour also entails that what men and women know is somewhat different, as each has gained expertise in areas that directly concern their areas of operation. As a consequence, the issue is not whether men have more knowledge than women but rather that what they know depends on their gender domains and social positions.

The fact that women lack mobility makes them more dependent on their male sons and husbands for certain information and knowledge. Hence, it would appear that by their very nature female-headed households are more liberal as compared to male-headed ones, as far as the flow of information within the households is concerned. Also in male-headed households information usually flows from the top downwards, whereas the flow is more lateral in female-headed ones.

Upon marriage women have to adopt their new family's way of doing things. As a result, the wife's knowledge is rarely taken into consideration when plans are being made. When she moves into her husband's home her knowledge and skills sometimes undergo a process of delegitimation or they simply lose their value. It is assumed that the woman can only know what she knows through her husband or other male relatives. Thus, even at school women take up those subjects that enable them to control what they know and to use their knowledge without any recriminations.

Having knowledge does not always mean using the knowledge. A variety of relationships affect what a person can and cannot do. Thus knowledge becomes an outcome of negotiation. As indicated in some of the discussions, women also dispute the label given to them by men that they do not have knowledge where the cultivation of certain cash crops like cotton is concerned. Sometimes attributions of knowledge were linked to power games when men found it in their favour to label women as having no knowledge on cash crops possibly to justify why they had to control income from these crops. Thus attributions of knowledge or lack of it might have little to do with the presence or lack of such knowledge in certain individuals but might be a discourse used to indicate or justify a variety of local relationships.

Although women are sometimes consulted for decisions, decision-making is heavily skewed in favour of men such that what women know is hardly ever taken into consideration.

8

Conclusion

The book has investigated the struggles, negotiations, contestations and accommodations that take place between actors during the production of knowledge. A consistent argument throughout has been that knowledge should be regarded as primarily social, and its production as a social process. The central aim has been to understand how farming and farming knowledge are embedded in the social, economic, political and cultural lives of actors. One of the recurrent themes in the investigative journey has been the central position of witchcraft, which urges us to fully understand and come to terms with the importance of witchcraft and withcraft accusations in agriculture.

Many lessons can be learned from this study. The first lesson is that agricultural practices can be understood in the context of non-agricultural domains (Hebinck and Ruben 1998). Non-agricultural practices can influence agricultural practices. As this book demonstrates, government policies, politics, religion, magic, wealth, health and agriculture are all connected at some level. Fairhead (1993: 199) reaches a similar conclusion in his study of the Bwisha farmers in Zaire. For him, a

> ... focus on technology helps isolate agriculture from the social context or put another way, the farmer from the person. Researchers who are permitted to examine agriculture in terms of sgricultural knowledge can maintain themselves in ignorance of the multitude of non-agricultural influences which influence agricultural practices.

The study also dispels the notion that the government is powerful enough to direct the production and dissemination of knowledge to the docile farmer through its various experts. While people acknowledged the existence of the government and its impact on their lives through its various instruments of control, they did not regard the government as an entity that had to be obeyed all the time. I noted in Chapter 1 that resettled people still took up jobs in the formal sector despite laws explicitly prohibiting this, and people have also resisted various state laws in both the colonial and post-colonial era that were imposed from above under the guise of the transfer of knowledge.

Theoretical pitfalls

I have attempted to avoid at least two theoretical pitfalls in this book. The first theoretical trap was the dichotomy between what is known in the classical sociological literature as 'scientific' and 'traditional' knowledge, or rather the distinction between 'modernity' and 'tradition'. This pitfall was avoided not only by denying the existence of these dichotomies but also by showing that they were irrelevant for explaining social and cultural phenomena in the Mupfurudzi resettlement area where I worked. Knowledge can never be regarded as modern or traditional, whereas the notion of 'traditional' implies that the knowledge so defined is static and resistant to change. In contrast, as shown in this book, knowledge is always in a state of flux. Even knowledge that has been passed on from generation to generation is continually reworked to suit existing conditions. For instance, as shown in Chapter 5, the practice of *chisi* was altered by villagers to suit the requirements and needs of their resettled communities, and information that was received from so-called scientific sources was often reworked to suit the farmers' needs, in line with what they believed to be true. For instance, although farmers adopted the use of certified seed and fertilisers, they would still send the seed to the lion spirits and to the prophets for their blessing during fertility rituals. Thus, as far as farmers are concerned, knowledge cannot be compartmentalised into two opposing types – 'traditional' or 'modern' – since it does not fit snugly into either category.

At the practical level, knowledge is neither scientific nor traditional but simply local. To capture this, I adopted the conceptual terms of 'localisation' and 're-localisation' that take into account how knowledge is reworked to suit local conditions and needs. We should not draw rigid distinctions between 'natural' and 'social', or between 'culture' and 'science'. In daily life these concepts are embedded within each other. Therefore, the scientist has to understand the farmer in his or her context, otherwise projects of high scientific quality may prove to be dismal failures shen it comes to implementation.

The second pitfall I tried to avoid was a conception of knowledge as a resource 'out there' waiting to be used. Knowledge is socially constructed and knowledge outcomes are often not consciously calculated or even intended by anyone. All knowledge is reworked to suit the available conditions, and the context in which the knowledge is applied. I have avoided this pitfall by showing that knowledge is a result of negotiations. Farmers did not regard information imparted by the so-called knowledge experts as simply knowledge. There were some rigorous tests that this knowledge had to undergo to be accepted as such. For instance, the knowledge bearer's symbolic capital often determined whether a knowledge claim was legitimised or not. Sometimes knowledge experts were suspected of trying to make money out of farmers' misfortunes, for example when they were recommending expensive drugs to cure livestock diseases. Scientifically, the experts might have been correct in recommending such drugs but because of the expense involved, this knowledge was viewed with suspicion by farmers who would seek solutions elsewhere. Sometimes, farmers carried out their own experiments to test the authenticity of the 'expert' knowledge claims, and in some cases (see Chapter 4) would arrive at conclusions that were far different from those of the experts.

Experts and farmers

As shown in Chapter 1, both colonial and post-independence governments developed ambitious knowledge projects for which they intended to recruit farmers: the farmer being the object that had to be acted upon. The farmer had to be supplied with knowledge and because he was regarded as perpetually 'ignorant', the only salvation could come from knowledge that was to be injected from outside. With the help of experts, governments attempted to introduce various mechanisms to ensure the farmer was schooled into the science of farming. Master Farmer training, field days, and village meetings were held, with experts leading the discussions and defining the topics.

The experts also adopted ways of classifying farmers, so that those who were considered not good enough would change their ways and model the 'good' farmers. As noted in the preceding chapters, these interventions had minimal impact on farmers' behaviour since they were recruited into 'projects' they had little interest in. Farmers had their own projects, which were different from the projects of government officials and other experts. Farmers were interested in reducing the cost of production, for instance they wanted to know how to produce better yields with low-cost seed, making minimum use of fertilisers. They were also concerned with finding low-cost and effective cures for animal diseases.

The lack of interest of farmers in the projects of the 'experts' was evidenced by the way the farmers tried to engage the experts and their resources into their own projects. Farmers know what they want. To be effective experts have to understand farmers' worldviews. Alternatively, just as farmers managed at times to hijack experts' projects to serve their own, so the experts should devise means of sometimes seizing farmers' projects to their own benefit. To maintain their relevance to resource-poor farmers, agricultural experts also have to focus on low-input farming with better returns, as well as on the use of locally available inputs for sustainable farming.

A recurrent theme throughout this book has been that knowledge is understood differently by different actors. This is not akin to reducing everything to interpretation, but acknowledges that the different life-worlds of the different actors within any social situation influence how they view certain available information and how they will eventually act. For example, those farmers who interpreted their poor yields by alluding to the evil works of people with bad magic never questioned their farming methods in the way scientists would do. This was so because their interpretation of poor harvests precluded the consideration of other factors, just like rigid scientism would rule out all magical or spiritual explanations. Depending on where actors are situated socially, they observe things and interpret them differently, so that the same thing can be attributed to different causal factors. Consequently, rural development workers must not be overly scientific, ignoring the different perceptions and meanings that people attach to activities and interventions, since this can bring about the failure of scientifically sound projects.

However, how people interpret certain events and occurrences often depends on larger frames of meaning and action. For instance, in issues of ill-health, people were quick to blame witches without investigating other probable causes such as the scientific origins of certain forms of pathology. When it comes to crop failure, witches may be accused of stealing crops, or experts suspected of knowingly supplying useless seed with the intention of discrediting the government. People may also point to angry ancestors who need to be appeased. However, as discussed in the previous chapters, such interpretations and meanings are largely dependent on the distribution of power and resources. Actors with differential access to power and resources can interpret the same event but come out with different conclusions.

Meanings are important to understand everyday life. It is the meanings that people attach to events, occurrences and deeds that give the impetus to take up certain actions. These meanings also lead people to either or not adopt certain things as knowledge. At least, at the local level, meanings attached to things or their properties lead to the production of certain forms of knowledge. The attached meanings give rise to theories. Even if these theories can be proved false by adopting scientific rationality, they are often pointers to behaviour, which also has implications for knowledge and its production. For instance, when farmers concluded that it was the bitterness of certain herbs and medicines that was responsible for their healing qualities, farmers experimented with a local beer brew that also had a bitter taste to try and cure animal diseases. The beer was found to be effective against certain animal diseases. Thus, although the meanings and their interpretation gave rise to a theory based on a false premise, the actions it produced had practical implications for knowledge and the rearing of animals.

For intervention to be effective, AREX and other experts should understand people's belief systems and the meanings they attach to certain things. As discussed in this book, the way people interpret and attach meanings to certain actions can often prevent them from adopting, or encouraging the adoption of, certain behaviour. For example, if people would decide that spraying food crops such as maize with modern chemicals to prevent attacks from pests was an affront to the spirits of the land, they would not use the spray, even if it could save their crops: something similar happened in the case of the army worm outbreak of 1995, when some people did not spray their crops. This was not because they did not know that spraying cabaryl 85 could at least spare some of their maize from the army worm attack but because they would rather get permission from the spirits first. It is very important for rural development workers to understand such cultural responses, because this will enable them to negotiate these meanings and beliefs with a view to changing people's behaviour without antagonising them.

Farmers strategise in their dealings with other actors and agents. With respect to experts, farmers did not follow them blindly and neither did they entirely disconnect themselves from them. They employed various linking and de-linking strategies in an attempt to maximise their gains from each encounter. During face-to-face encounters between the resettled farmers and outside institutions and agents farmers were not always powerless. For instance, if there was mistrust between the farmers and experts, and if farmers were not consulted, decisions by policy-makers could be frustrated at the implementation level. For example, when the government took a unilateral decision without consultation to force farmers to brand their cattle to protect the cattle from theft, the farmers resisted such a move. The farmers thought that selling the branding equipment was government's way to get money from them, since each farmer had to have his own unique registered brand. Farmers resisted this move and the plan was never implemented.

In order to succeed, experts need to consult with farmers at every level of policy-making, not just at the implementation level, as is often the case. The traditional stereotype of the peasant being resistant to change precludes experts questioning their own ability and conduct when certain projects fail. It is too easy to just blame the farmer. In my view, farmers are not resistant to change but they resist projects that they believe do not serve their interests well. Before people can adopt new technology, it must be something they can go along with and find both acceptable and useful. This has wider implications for technology and its use. Experts should understand that technology is not value-free, an artefact to be used, but instead needs to be interpreted and understood in a social context which will determine its success or failure.

Knowledge is not always what it seems and is not always positive. Knowledge can be disempowering to the one who is equipped with it. Although modern scientific knowledge is very efficient, it has made farmers more dependent on agro-business as opposed to the independence they enjoyed when all the resources were locally available. Not only scientific knowledge could be disempowering to the farmer, but also local knowledge. Thus although its value has to be recognised, local knowledge should not be regarded as the solution to all farming problems as it is not always effective. For example, although farmers liked to use local seed as it entailed less input from outside and reduced production costs, they recognised that people who used certified seed and fertilisers generally got higher crop yields than those who used local seed and manure.

An issue that needs to be constantly emphasised is that knowledge and technologies are not only technical but also social in nature. Knowledge can be scientific but, as discussed in the preceding chapters, it can also be recast by resorting to African systems of thought. This being the case, experts need to understand these African systems of thought if they are to have any lasting impact or indeed any impact at all.

Power is always contested and negotiated. Peasants are not always powerless and dependent, neither is the state always powerful and dominant. Experts had to tread carefully in their dealings with the farmers because if they acted in an overbearing and arrogant way they risked being branded as belonging to the political opposition that wanted to discredit the government. This threat in itself was a powerful weapon in the hands of farmers to keep experts in check, so that the experts could not even deal harshly with loan payment defaulters. The worst they could do was to deny the defaulter access to further loans, but they could not seize the defaulter's property to recover losses. The ability to tread the delicate ground between knowledge, politics and economics is an essential tool for the rural development worker, or else his or her effectiveness will be limited.

When their association with the experts could bring them advantages such as access to seed and fertiliser, farmers would attend agricultural meetings and classes convened by such experts. They would attend not because they expected to learn anything new, but simply to maintain good ties with the experts so that they would not deny them access to resources. On the other hand, experts could always threaten to withdraw seed and fertilisers to bring recalcitrant farmers to heel. This indicates that power is always contested and negotiated, and that farmers are not always powerless and dependent, nor is the state always powerful and dominant.

The fragmentary and contradictory nature of knowledge allowed farmers to manoeuvre within their social system and to negotiate to their advantage. This fragmentary nature of knowledge allowed people to work with a multiplicity of understandings, beliefs and commitments. It enabled farmers to bridge the gap between external and local knowledge. Farmers knew that to have good yields they needed good seed. However, there were different understandings of what good seed was. The scientific understanding was that good seed was properly and scientifically engineered to resist pests and diseases. Subscribing to this view, farmers bought commercial seed varieties if they could afford it, or else got the seed on loan if they were still eligible. However, the farmers' conception of good seed did not stop at certified seed but extended to include the importance of spiritual blessings to make the seed even better. Although for scientists good seed had nothing to do with priests and spirits but only with proper breeding, farmers were able to bridge that gap because for them knowledge did not have those essentialising qualities that set it apart as something distinct. For farmers, knowledge was contradictory and fragmentary which made it easy for them to believe in one thing at one time and maybe in a totally different thing at another time. Thus, to share Long's sentiment, one cannot distinguish between different forms of knowledge, but knowledge should be regarded as an outcome of the negotiations that take place between actors and their life worlds.

Heterogeneity

It has been a recurrent theme in this book that local farmers are not homogeneous in their outlook and neither are experts. This heterogeneity among local farmers means that expert policies and programmes are experienced differently, leading to diverse interpretations and actions. The heterogeneity of local farmers was also a result of the resettlement programme that brought together people from diverse backgrounds and communities who had varied reasons to apply for resettlement. Experts also differed in the way they implemented their knowledge, as was shown Chapter 6. Some officers were overbearing in their manner whereas others were engaging and persuasive in disseminating knowledge to farmers.

The discussion of gender in Chapter 7 argued that it is indeed a fallacy to say that because people in the same house practise the same things, they know the same things. It was shown in this chapter that upon marriage women had to adopt their new family's way of doing things. Consequently, merely observing what people do or how they cultivate their crops or rear their animals does not tell us what these people know. When it came to women in male-headed households, the link between practice and knowledge often became tenuous. For instance, the woman who had attended AREX lessons and acquired new knowledge could not implement what she had learned because her husband was resistant to the ideas. This woman continued with the practices that were now inimical to her new-found knowledge. Social powerlessness among women determined what women practised compared with what they knew. Thus, the relevant question to ask would not have been whether women and men had access to the same information but whether they both had equal chances to practise what they knew. Although there were no physical barriers or laws discriminating against women's access to certain kinds of information, women usually did not actively seek this information. This was mostly because they would not be able to implement the acquired knowledge if it was met with resistance from their husbands and sons. This I would regard as the interface between knowledge and culture. Although some women had knowledge, this knowledge would not get the recognition it deserved from their male guardians. If knowledge is not recognised it cannot be applied. The different forms of prevailing social relationships mediate the status of knowledge.

In concluding this study, let me borrow a statement from Pool (1994: 52): who says, 'I can never attain a final interpretation of what my informants "really" meant because there is no final interpretation'. My understanding and interpretation of the ethnographic case material has shown that indeed knowledge is not an artefact out there waiting to be used but is manufactured and produced as a result of struggles, negotiations, contestations and accommodations between farmers and outside agencies and institutions. These processes also take place among the individual farmers and families of the resettlement community where the research took place.

References

AGRAWAL, A. (1995) Indigenous and scientific knowledge: Some critical comments, *Indigenous knowledge monitor*, 3(3).

ALEXANDER, J. (2003) 'Squatters', veterans and the state in Zimbabwe, in Hammar, A., B. Raftapolous and S. Jensen (eds.) *Zimbabwe's unfinished business: Rethinking land, state and nation in the context of crisis*, Harare: Weaver Press, pp. 83-117.

ALVORD, A. (1958) *The development of native agriculture and land tenure in Southern Rhodesia*, unpublished.

ANDERSSON, J.A. (2002) *Going places, staying home. Rural urban connections in Buhera District, Zimbabwe*, Wageningen: PhD Thesis.

ASCHWANDEN, H. (1989) *Karanga mythology. An analysis of the consciousness of the Karanga in Zimbabwe*, Zimbabwe: Mambo Press.

ARCE, A. and N. LONG (1992) The dynamics of knowledge: Interfaces between bureaucrats and peasants, in Long, N. and A. Long (eds.) *Battlefields of knowledge: The interlocking of theory and practice in social research and development*, London: Routledge

ARCE, A. and N. LONG (2000) Reconfiguring modernity and development from an anthropological perspective, in Arce, A. and N. Long (eds.) *Anthropology and modernities: Exploring discourses, counter tendencies and violence*, London: Routledge.

ARCE, A. and E. FISHER (2003) Knowledge interfaces and practices of negotiation: Cases from a women's group in Bolivia and an oil refinery in Wales, in Pottier, J., A. Bicker and P. Sillitoe (eds.) *Negotiating local knowledge: Power and identity in development*, London: Pluto Press, pp. 74-97.

ASHFORTH, A. (1998) Reflections on spiritual insecurity in a modern African City (Soweto), *African Studies Review*, 41(3): 39-67.

BARBALET, J.M. (1985) Power and resistance, *British Journal of Sociology*, 36(4): 531-548.

BARNES, B., D. BLOOR and J. HENRY (1996) *Scientific knowledge: A sociological analysis*, London: Athlone press.

BARTH, F. (2002) An anthropology of knowledge, *Current Anthropology*, 43(1): 1-18, full text version: www.journals.uchicago.edu/CA/journal.

BRATTON, M. (1987) The comrades and the countryside: The politics of agrarian policy in Zimbabwe, *World politics*, 39(2): 174-202.

BRATTON, M. (1980) *The local politics of rural development: Peasant and party state in Zambia*, Hanover: University Press of New England.

BEACH, D.N. (1979) 'Chimurenga': The Shona rising of 1896-97, *The journal of African History*, 20(3): 395-420.

BERGER, P.L and T. LUCKMANN (1971) *The social construction of reality. A treatise in the sociology of knowledge*, Great Britain: Penguin.

BESHAH, T. (2003) *Understanding farmers explaining soil and water conservation in Konso, Wolaita and Wello, Ethiopia*, unpublished thesis.

BLICK, G.A.B. (1972) *The history of Shamva*, National archives of Zimbabwe.

BOLLEY, H.L. (1892) The status of experimental agriculture, *Science*, 19 (484): 270-271.

BOLDING, A. (2004) *In hot water. A study on sociotechnical intervention models and practices of water use in smallholder agriculture, Nyanyadzi catchment, Zimbabwe*, Wageningen.

BOURDIEU, P. (1990) *The logic of practice*, London: Polity Press.

BOURDIEU, P. (1990) *In other words. Essays towards a reflexive sociology*, London: Polity Press.

BOURDILLON, M., P. HEBINCK, J. HODDINOTT, B. KINSEY, J. MARONDO, N. MUDEGE AND T. OWENS (2002) *Impact of agricultural research on poverty reduction case study 3: Assessing the impact of HYV maize in resettlement areas of Zimbabwe*, Washington DC: International Food Policy Research Institute.

BOURDILLON, M.F.C. (1990) *Religion and society. A text for Africa*, Zimbabwe: Mambo Press.

BOURDILLON, M.F.C. (1989) Medicines and symbols, *Zambezia*, 16(1): 29-44.

BOURDILLON, M.F.C. (1993) *Where are the ancestors: Changing culture in Zimbabwe*, Harare: UZ.

BOURDILLON, M.F.C. (1987) *The Shona peoples. An ethnography of the contemporary Shona with special reference to their religion* (third revised edition), Zimbabwe: Mambo Press.

BOURDILLON, M.F.C. (1987) Guns and rain taking structural analysis too far, review article, *Africa*, 57(2): 263-274.

BOURDILLON, M.F.C. (1984) Religious symbols and political change, *Zambezia*, XII: 39-55.

BREWER, J.D. (2000) *Ethnographic understanding*, Social research series, London: Bryman.

BUSH, R. and L. CILFFE (1984) Agrarian policy in migrant labour societies: Reform or transformation, review, *Review of African Political Economy*, 29.

CARGILL (2004) *www.cargill.com/worldwide/zimbabwe.*

CARNEY, J. (1993) Converting wet lands, engendering the environment: The intersection of change and agrarian changes in the Gambia, *Economic Geography*, 62(4): 329-348.

CALLON, M. (1986) The Sociology of an actor-network: the case of the electric vehicle, in Callon, M., J. Law and A. Rip (eds.) *Mapping the dynamics of science and technology*, London: MacMillan Press.

CALLON, M. (1992) The Dynamics of techno-economic networks, in Coombs, R. *et al.* (ed.) *Technological change and company strategies: economic and sociological perspectives.* London: Harcourt.

CALLON, M., and J. LAW (1986) On the construction of sociotechnical networks: Content and context revisited, *Sociology of Science Past and Present*, 8: 57-83.

CALLON, M., and LATOUR, B. (1992) Don't throw the baby out with the bath school. A reply to Collins and Yearly, in Pickering, A. (ed.) *Science as practice and culture. From science as knowledge to science as practice*, Chicago: University of Chicago Press.

CHABAL, P. (1996) The African crisis: Context and interpretation, in Werbner, R. and T. Ranger (eds.) *Post colonial identities in Africa. Post colonial encounters series*, London: Zed Books, pp. 29-54.

CHAKWERA, N. (2002) *Agribank in Zimbabwe - Agricultural Development and Assistance Fund (ADAF), Technical services, Agricultural Bank of Zimbabwe, part II, Emerging rural project finance models*, available online: http://www.zoic.co.zw.

CHAMBERS, R. (1997) *Whose reality counts? Putting the first last*, London: Intermediate Technology Publications.

CHAMBERS, R. (1983) *Rural development: Putting the last first*, London: Longman.

CHAVHUNDUKA, G. (1980) Realities of Witchcraft, *Zambezia*, 13(2): 130-5.

CHAVHUNDUKA, G. (1986) Witchcraft and the Law in Zimbabwe, *Zambezia*, 8(2): 129-47.

CHEATER, A.P. and R.B. GAIDZANWA (1996) Citizenship in neo-patrilineal states: Gender and mobility in southern Africa, *Journal of Southern African Studies*, 2(2): 189-200.

CHEATER, A.P. (1984) *Idioms of accumulation: Rural development and class formation among free holders in Zimbabwe*, Gweru: Mambo Press.

CHITSIKE, F. (2003) *A critical analysis of the land reform programme in Zimbabwe*, Second FIG regional conference, Marrakech, Morocco, December 2-5.

CHIVAURA, V., and C. MARARIKE (eds.) (1998) *The human factor approach to development in Africa*, Harare: University of Zimbabwe.

CIAT (2005) *Ciat in Synthesis. Solutions that cross frontiers*, available online: http://www.ciat.cgiar.org.

CIEKAWY, D. and P. GESCHIERE (XXXX) Containg witchcraft: Conflicting scenarios in postcolonial Africa, *Africa Studies review*, 41(3): 1-14.

CLIFFORD, J. (1988) *The predicament of culture: twentieth - century ethnography, literature, and art*, Cambridge: Cambridge University Press.

CORTESE, A.J. (1995) The rise, hegemony and decline of the Chicago School of Sociology, 1892-1945, *The Social Science Journal*, 32(3): 235-54.

COTTON COMPANY (2004), *www.thecottoncompany.com.*

DANEEL, M.L. (1971) *Old and new in Southern Shona independent churches, Vol. 1, Background and rise of the major movements*, The Hague: Mouton.

DANEEL, M.L. (1987) *Quest for belonging. Introduction to a study of African independent churches*, Gweru: Mambo Press.

DEKKER, M. (2004) *Risk, resettlement and relations: Social security in rural Zimbabwe*, Vrije Universiteit of Amsterdam.

DEWALT, K.M., B.R. DEWALT and C.B. WAYLAND (1998) Participant Observation, in Bernard, R.H. (ed.) *Handbook of methods in cultural anthropology*, London: Altamira Press.

DILLON-MALONE, C.M. (1978) *The Korsten basket makers. A study of the Masowe apostles. An indigenous African religious movement*, University of Manchester.

DRINKWATER, M. (1994) Knowledge, consciousness and prejudice: Adaptive agricultural research in Zambia, in Scoones, I. and J. Thompson (eds.) *Beyond farmer first. Rural people's knowledge, agricultural research and extension*, UK: Intermediate Technology Publications.

DOLAN, C.S. (1997) Gender and witchcraft in agrarian transition: The case of Kenyan horticulture, *Development and Change*, 33(4): 659-681.

EHRENREICH, B. and D. ENGLISH (1973) *Complaints and disorders: The sexual politics of sickness*, UK: Writers and Readers Publishing Cooperative.

EVANS-PRITCHARD, A. (1937) *Witchcraft oracles and magic among the Azande*, Oxford: Clarendon Press.

FAIRHEAD, J. (1993) Representing knowledge. The 'new farmer' in research fashions, in Pottier, J (ed.) *Practising development: Social science perspectives*, London: Routledge.

FEATHERSTONE, M. (1990) Global culture: An introduction, *Theory, culture and society*, 7(2 and 3).

FERGUSON, A.E. (1994) Gendered science: A critique of agricultural development. *American anthropologist New series*, 96(3): 540-52.

FISIY, C., and P. GESCHIERE (1996) Witchcraft, violence and identity: different trajectories in post-colonial Cameroon, in Werbner, R., and t. Ranger (eds.) *Post colonial identities in Africa. Post colonial encounters series*, London, pp. 193-221.

FREIRE, P. (1993) *Pedagogy of the oppressed* (new revised edition translated by M.B. Ramos), London: Penguin.

FOUCAULT, M. (1967) *The archaeology of knowledge*, London: Routledge.

GAIDZANWA, R.B. (1995) Land and the economic empowerment of women: A gendered analysis, *Southern African Feminist Review*, 1(1): 1-12.

GEERTZ, C. (1998) *Works and lives; The anthropologist as author*, Cambridge: Polity Press.

GELFAND, M. (1956) *Medicine and magic of the mashona*, Cape Town: Juta and Company Limited.

GESCHIERE, P. (1997) *The modernity of witchcraft: Politics and the occult in postcolonial Africa*, Virginia (USA): University Press of Virginia.

GOEBEL, A. (1999) 'Here it is our land – the two of us': Women, men and land in a Zimbabwean resettlement area, *Journal of Contemporary African Studies*, 17(1) 75-96.

GOHEEN, M. (1996) *Men own fields, women own the crops: Gender and power in the Cameroon grassfields*, Wisconsin (USA): University of Wisconsin.

GOLINSKY, J. (1998) *Making natural knowledge constructivism and the history of science*, Cambridge (USA): Cambridge University Press.

GOODY, J. (1977) *The domestication of the savage mind*, Cambridge (UK): Cambridge University Press.

GREEN, R.H. and S.H. HYMER (1966) Cocoa in the Gold Coast. A study in the relations between African farmers and Agricultural experts, *The Journal of Economic History*, 26(3): 219-319.

GUNNING, H.W., J. HODDINOT, B. KINSEY and T. OWENS (2000) Revisited, forever gained: Income dynamics in the resettlement area (version February 2000), available online: www.eudnet.net.

GMB (2002) *www.dexelzim.co.zw/gmb/html.*

HAGMANN, J., E. CHUMA, M. CONNOLLY, and K. MURWIRA (1997) Propelling change from bottom up: Institutional reform in Zimbabwe, in International Institute for Environment and Development, *Gate keepers Series 7*, available online: www.iied.org/docs.

HAGENAARS, A., and K. DE VOS (1988) The definition and measurement of poverty, *The journal of human resources*, 23(2): 211-21.

HAMMERSLEY, M. (1998) *Reading ethnographic research*, London: Longman.

HANNERZ, U. (1990) Cosmopolitans and locals in wolrd culture, *Theory, Culture and Society*, 7(2 and 3).

HAVERKORT, B., K. VAN 'T HOOFT and W. HIEMSTRA (eds.) (2003) *Ancient roots, new shoots: Endogenous development practice*, London: Zed Books.

HAVERKORT, B., D. MILLAR and C. GONESE (2003) Knowledge and belief systems in Sub-Saharan Africa, in Haverkort, B., K. van 't Hooft and W. Hiemstra (eds.) (2003) *Ancient roots, new shoots: Endogenous development practice*, London: Zed Books.

HEBINCK, P., and M.F.C. BOURDILLON (eds.) (2001) *Women, men and work. Rural livelihoods in south-eastern Zimbabwe*, Harare: Weaver Press.

HEDRICK, W.O. (1918) The tutored farmer, *The Scientific Monthly*, 7(2): 158-65.

HELLEINER, G.K. (1970) Development policies for Africa in the 1970s, *Canadian Journal of African Studies*, 4(3): 285-303.

HERBST, J. (1990) *State politics in Zimbabwe*, Harare: University of Zimbabwe Press.

HIRST, P. (1985) Power/ knowledge – Constructed space and the subject, in Fardon, R. (ed.) *Power and knowledge. Anthropological and sociological approaches*, Edinburgh: Scottish Academic Press.

HOWARD-BORJAS, P. (1999) Some implications of gender relations for plant genetic resource management, *Biotechnology and Development Monitor*, 37: 2-5.

HULME, D., and M. TURNER (1990) *Sociology and development. Theories, policies and practices*, New York: Harvester Wheatsheaf.

INTERNATIONAL LABOUR ORGANISATION (1995) *Gender poverty and employment. Turning capabilities into entitlements*, ILO.

JACKSON, M. (1989) *Paths toward a clearing*, Bloomington: Bloomington University Press.

JACOBS, S. (1993) Women and land resettlement in Zimbabwe, *Review of African Political Economy*, 23: 33-50.

JACOBS, S. (1991) Land resettlement and gender in Zimbabwe: Some findings, *Journal of Modern African Studies*, 29(3): 521-8.

JACOBS, S. (1990) Zimbabwe: State, class, and gendered models of land resettlement, in Parpart, J.L. and K.A. Staudt (eds.) *Women and the state in Africa*, London: Lynne Reinner Publishers, pp. 161-184.

JONES, W.O. (1987) Food crop marketing boards in tropical Africa, *The journal of modern African Studies*, 25(3): 375-402.

KEESING, M.R. (1987) Anthropology as interpretive quest, *Current Anthropology*, 28(2): 161-76.

KINSEY, B. (1999) Land reform, growth and equity: Emerging evidence from Zimbabwe's resettlement programme, *Journal of Southern African Studies*, 25(2): 173-96.

KINSEY, B. (1982) Forever gained: Resettlement and land policy in the context of national development in Zimbabwe, *Africa*, 52: 92-113.

KNORR-CETINA, K.D. (1983) The ethnographic study of scientific work: Towards a constructivist interpretation of science, in Knorr-Cetina, K.D. and M. Mulkay (eds.) *Science observed. Perspectives on the social study of science*, London: Sage.

KOROVIKIN, I. (1986) Oral contraceptives in a southern Italian community, *Current Anthropology*, 27(1):80-2.

KWANGWARI, C. (forthcoming) *An analysis of women's participation in formal and non-formal politics at the district and subdistrict levels: The case of Goromonzi*, UZ, Sociology: MPhil. Thesis.

LAN, D. (1985) *Guns and rain guerrillas and spirit mediums in Zimbabwe*, London: James Currey.

LATOUR, B. (1983) Give me a laboratory and I will raise the world, in Knorr-Cetina, K.D. and M. Mulkay (eds.) *Science observed. Perspectives on the social study of science*, London: Sage.

LATOUR, B. (1993) *We have never been modern*, New York: Harvester Wheatsheaf.

LEEUWIS, C. (2004) (third edition) *Communication for rural innovation: Rethinking agricultural extension*, UK: Blackwell Science.

LETT, J. (1991) Interpretive anthropology: Metaphysics and the paranormal, *Journal of Anthropological Research*, 47.

LEVY, R.I. and D.W. Hollan (1998) Person-centered interviewing and observation, in Bernard, R.H. (ed.) *Handbook of methods in cultural anthropology*, London: Altamira Press.

LITTLETON, C.H. (1965) A two dimensional scheme for the classification of narratives, *The Journal of American Folklore*, 78(307): 21-7.

LOGAN, E.A. (1985) The Shamva story, *Heritage*, 5: 17-30.

LONG, N. (2006, forthcoming) Resistance, agency and counterwork: A theoretical positioning, in Wright, W and G. Middendorf (eds.) *Food fights*, Penn State University Press.

LONG, N. (2004) Contesting policy ideas from below, in Boäs, M. and D. McNeill (eds.) *Global institutions and development: Framing the world?*, London and New York: Routledge, pp. 24-40.

LONG, N. and A. LONG (1992) *Battlefields of knowledge. The interlocking of theory and practice in social research and development*, London: Routledge

LONG, N. (1996) Globalisation and localisation: New challenges to rural research, in Moore, H.L. (ed.) *The future of anthropological knowledge*, London: Routledge, pp. 37-59.

LONG, N. (1992) From paradigm lost to paradigm regained. The case for an actor oriented sociology of development, in Long, N., and A. Long (eds.) *Battlefields of knowledge. The interlocking theory and practice in social research and development*, London: Routledge.

LONG, N., and M. VILLAREAL (1994) The interweaving of knowledge and power in development discourse, in Scoones, I. and J. Thompson (eds.) *Beyond farmer first. Rural people's knowledge, agricultural research and extension*, UK: Intermediate Technology Publications

LONG, N., and M. VILLAREAL (1993) Exploring development interfaces from the transfer of knowledge to the transfer if meaning, in Schuurman, F.J. (ed.) *Beyond the impasse: New directions in development theory*, London: Zed Books.

LONG, N. (2001) *Development sociology actor perspectives*, London and New York: Routledge.

LONG, N. and J.D. VAN DER PLOEG (1994) Heterogeneity, actors and structure: Towards a reconstruction of the concept of structure, in Booth. D. (ed.) *Rethinking social development: Theory and practice*, London: Harlow Longman

MANGO, N.A.R. (2002) *Husbanding the land. Agricultural development and social and technological change in Luoland Kenya*, Wageningen University, Thesis.

MANGO, N. and P. HEBINCK (2004) Cultural repertoires and socio-technological regimes: Maize in Luoland, in Wiskerke, J.C. and J.D. van der Ploeg *Seeds of transition: Essays on novelty production niches and regimes in Africa*, Assen: Royal Van Gorcum.

MANZUNGU, E. and P. VAN DER ZAAG (eds.) (1996) *The practice of small holder irrigation: Case studies from Zimbabwe*, Harare: University of Zimbabwe.

MARARIKE, C.G. (1999) *Survival strategies in rural Zimbabwe: The role of assets, indigenous knowledge and organisation*, Harare: Mond Press.

MATE, R. (2001) Land women and sugar in Chipiwa, in Hebinck, P. and M.F.C. Bourdillon (eds.) *Women, men and work: Rural livelihoods in south-eastern Zimbabwe*, Harare: Weaver Press

MATOSE, F. and B. MUKAMURI (1993) Rural people's knowledge and extension practice: Trees, people and communities in Zimbabwe's communal lands: Towards a theoretical framework, *Sustainable Agriculture Programme Research Series*, 1(1).

MATOSE, F. (1997) Conflicts around forest reserves in Zimbabwe, in Leach, M., R. Mearns and I. Scoones (eds.) *Community based sustainable development: Consensus or conflict, IDS bulletin*, 28(4), pp. 69-78.

MATOWANYIKA, J.Z.Z. (1991) *Indigenous resources management and sustainability in rural Zimbabwe: An exploration of practices and concepts in Common land*, Ontario, Canada: Waterloo.

MBIBA, B. (1999) Security of tenure, development victims, and the limits of environmental impact assessment in Zimbabwe's Communal Lands, *Development in Practice*, 9(3): 316-22.

MAURYA, D.M. (1989) The innovative approach of Indian farmers, in Chambers, R., A. Pacey and L.A. Thrupp (eds.) *Beyond farmer first*, Intermediate Technology Publication.

MCCULLOUGH, A. (1987) The truth of historical narratives, *History and Theory*, 26(4): 30-46.

MITCHEL, C. (1983) Case and situation analysis, *Sociological Review*, 31: 187-212.

MITCHELL, T. (1990) Everyday metaphors of power, *Theory and Society*, 19(5): 545-77.

MOMBESHORA, S. (1994) Witches, witchcraft and the question of order: A view from a Bena village in the Southern highlands, in Abrahams, R., (ed.) *Witchcraft in contemporary Tanzania*, Cambridge African Monographs, African Studies Centre, Cambridge (UK):Cambridge University Press.

MOMBESHORA, S. (1998) Local institutional adaptations to the economic reform programme, in Mombeshora, S. (ed.) *Research in the social impacts of economic reform programme on rural Zimbabwe*, Harare: Mondi Books

MONU, E.D. (1982) Improving agricultural practices among African small holders - The contribution of adoption and diffusion of innovation to agricultural development in Africa, *African Studies Review*, 25(4): 117-26.

MOORE, D.S. (1998) Clear waters and muddied histories: Environmental history and the politics of community in Zimbabwe's Eastern Highlands, *Journal of Southern African Studies*, 24(2): 377-403.

MOORE, D.S. (1993) Contesting terrain in Zimbabwe's eastern highlands: Politics, ecology, ethnography and peasant resource struggles, *Economic Geography*, 69(4): 380-401.

MOYO, S. (2004) *The land and agrarian question in Zimbabwe*, paper presented at the Conference on The Agrarian Constraint and Poverty Reduction: Macroeconomic Lessons for Africa, Addis Ababa, 17-18 December.

MOYO, S. (2000) The political economy of land acquisition and redistribution in Zimbabwe 1990-1999, *The Journal of Southern African Studies*, 26(1): 5-28.

MOYO, S. (1998) *The land acquisition process in Zimbabwe (1997/98)*, United Nations Development Programme (UNDP).

MOYO, S. (1996) A gendered perspective on the land question in Zimbabwe, *Southern African Feminist Review (SAFERE)*, 1(1).

MUNDY, P.A. and J.L. COMPTON (1995) Indigenous communication and indigenous knowledge, in Warren, D.M. *et al.* (eds.) *The cultural dimensions of development. Indigenous knowledge systems*, London: Intermediate Technology.

MURPHREE, M.W. (1971) Religious interdependency among the Budjga Vapostori, in Barrett, D.B. (ed.) *African initiatives in religion*, Nairobi: East African Publishing House.

MURWIRA, K., J. HAGMANN AND E. CHUMA (2001) Mainstreaming participatory approaches to SWC in Zimbabwe, in Reij, C. and A. Waters-Bayer (eds.) *Farmer innovation in Africa. A source of inspiration for agricultural development*, London: Earthscan, pp. 300-09.

MUSHUNJE, M.T. (2001) *Women's land rights in Zimbabwe*, Madison: BASIS Land tenure Center, available online: http://www.wisc.edu/1tc/basis.html.

MUSONI, F. (1999) *The impact of human activities on the environment. Study of environmental degradation in Buhera District from the late pre-colonial period to the 1950s* BA Thesis, University of Zimbabwe.

MUTANGADURA, G. (1997) *Meeting development objectives with agricultural research priority setting in Zimbabwe*, PhD Thesis, Agricultural Economics, Blackburg, Virginia, available online: http//scholar.lib.vt.edu/thesis.

NIEHAUS, I., E. MOHLALA and K. SHOKANE (2001) *Witchcraft, power and politics: Exploring the occult in the South African Lowveld*, London: Pluto Press.

NIEHAUS, I.A. (1998) The ANC's dilemma: The symbolic politics of three witch hunts in the South African Lowveld 1990-1995, *African Studies Review*, 41(3): 93-118.

ONSELEN, C. van (1972) The reaction to rinderpest in southern Africa 1896-1897, *The Journal of African History* ,13(3): 473-88.

PALMER, R. (1977) The agricultural history of Rhodesia, in Palmer, R. and N. Parsons *The roots of rural poverty in central and southern Africa*, London: Heinemann, pp. 221-54.

PARKIN, D. (1985) Controlling the u-turn of knowledge, in Fardon, R. (ed.) *Power and knowledge. Anthropological and sociological approaches*, Edinburgh: Scottish Academic Press.

PARPART, J.L. and K.A. Staudt (1990) *Women and the state in Africa*, London: Lynne Reinner Publishers, pp. 1-19.

PHIMISTER, I. (1977) Peasant production and underdevelopment in Southern Rhodesia, 1890-1914, with particular reference to the Victoria district, in Palmer, R. and N. Parsons (eds.) *The roots of rural poverty in central and southern Africa*, London: Heinemann, pp. 255-67.

PLOEG, D.J. van der (2003) *The virtual farmer: Past, present and future of the Dutch peasantry*, Netherlands: Gorcum.

POOL, R. (1994) *Dialogue and the interpretations of illness: Conversations in a Cameroonian village*, London: Berg Publishers.

POTTIER, J. (2003) Negotiating local knowledge: An introduction, in Pottier, J., A. Bicker and P. Sillitoe (eds.) *Negotiating local knowledge: Power and identity in development*, London: Pluto Press, pp. 1-29.

POULTON, C., R. DAVIES and I. MATSHE (2002) *A review of Zimbabwe's Agricultural economic policies 1980-2000*, available online: www.imperial.ac.uk/agriculturalsciences.

POVERTY ASSESSMENT SURVEY (1995) Ministry of Public Service and Social Welfare (CSO).

PROBST, P. (1999) Mchape '95 or, the sudden fame of Billy Goodson Chisupe: Healing social memory and the enigma of the public sphere in post Banda Malawi, *Africa*, 69: 1, 110-130.

RANGER, T. (1967) *Revolt in Southern Rhodesia 1896-7: A study of African Resistance*, Evanston: Northwestern University Press

RASMUSSEN, S.J. (2004) Reflections on witchcraft, danger, and modernity among the Tuareg, *Africa*, 74(3): 314-15.

RHOADES, R. (1989) The role of farmers in the creation of Agricultural technology, in Chambers, R., A. Pacey and L.A. Thrupp (eds.) *Beyond farmer first*, Intermediate Technology Publication.

RICHARDS, P. (1985) *Indigenous agricultural revolution*, Boulder: Westview Press.

ROBERTS, R. (2000) History and memory: The power of statist narratives, *International Journal of African Studies*, 33(3): 513-22.

ROBERTSON, A.F. (1991) *Beyond the family: The sociological organisation of human reproduction*, UK: Polity Press

ROOSER, C. and C. HARRIS (1983) *The family and social change: A study of family and kinship in a South Wales town*, students' edition, London: Routledge and Kegan Paul.

ROSALDO, R.I. (1997) A note on Geertz as cultural essayist, *Representations*, 59(summer): 30-4.

ROTH, G. (2001) The position of farmers' local knowledge within agriculture extension, research and development cooperation, *Indigenous Knowledge and Development Monitor*, 9(3): 10-12.

RUBEN, R. and P. Hebinck (1998) *Rural livelihoods and strategies: Straddling farm and non-farm activities. 15th International Symposium on Farming systems Research and Extension*, South Africa, Pretoria, 29 November-4 December 1998.

RUKUNI, M. (1994) *Zimbabwe: Report of the Commission of Enquiry into appropriate agricultural land tenure systems*, Harare: Government Printer.

RUTHERFORD, B. (1999) To find an African witch Anthropology modernity and witch finding in Zimbabwe, *Critique of Anthropology*, 1: 89-109.

SADOMBA, W.Z. (1999a) *The impact of settler colonialism on indigenous agricultural knowledge in Zimbabwe: Fusion, confusion or negation*, Wageningen, Msc Thesis.

SADOMBA, W.Z. (1992) *The History of land management in Zimbawe: An exposition of the impact of settler colonialation on african land philosophy*, Centre for Indigenous Cultures in Southern Africa (CISCA).

SADOMBA, W.Z. (1999b) *Changing agro-ecological religion and changing agronomic practices: African cosmology and women versus western scientific agriculture in Zimbawe*, paper presented at the International Conference on Women and the Environment, Harare.

SAFIRE (2002) Chivi food security project. A process approach. Thematic area: Sustainable agriculture, food safety and food security, *Southern Alliance for Indigenous Resources*.

SCHUTZ, A. (1964) *Collected papers 1: The problem of social reality* (edited and Introduced by M. Natanson, preface by H.L. van Breda), The Hague: Martinus Nijhoff.

SCHUTZ, Th.W. (1964) *Transforming traditional agriculture*, New Haven: Yale University Press.

SCOONES, I. and J. THOMPSON (1993) Overview papers, in Scoones, I. and J. Thompson (eds.) *Beyond farmer first - Rural people's knowledge, agricultural research and extension practice*, Towards a theoretical Framework in Sustainable Agriculture, Programme Research Series 1(1), UK: Intermediate Technology Publications.

SCOONES, I. and J. THOMPSON (1994) (eds.) *Beyond farmer first. Rural people's knowledge, agricultural research and extension*, UK: Intermediate Technology Publications.

SCOONES, I. and B. COUSINS (1989) A participatory model of agricultural research and extension: The case of *vleis*, trees and grazing schemes in the dry south of Zimbabwe, *Zambezia*, XVI(i): 45-65.

SEVERAL AUTHORS (1936) Cooperation in large scale experimentation, supplement to the *Journal of the Royal Statistical Society*, 3(2): 115-36.

SHAMBARE, M. and M.F.C. BOURDILLON (2002) *Gossip and social control*, Revisited report to the Centre for the Study of African Economies, 25 March: 13.

SHEPHERD, A. (1998) *Sustainable rural development*, London: MacMillan Press.

SHUTT, A.K. (1997) Purchase area farmers and the middle class of southern Rhodesia, c. 1931-1952, *The International Journal of African Historical Studies*, 30(3) 555-81.

SKALNES, T. (1995) *The politics of economic reform in Zimbabwe. Continuity and change in development*, London: Macmillan Press.

SEED CO (2004) *Seed manual*, available online: www.seedco.zw.

SEN, A. (1980) *Poverty and famines: An essay on entitlement and deprivation*, Oxford, Clarendon Press

SEUR, H. (1992) The engagement of researchers and local actors in the construction of case studies and research themes: Exploring methods of restudy, in Long, N. and A. Long (eds.) *Battle fields of knowledge: The interlocking of theory and practice in social research and development*, London: Routledge.

SILLITOE, P. (1998) The development of indigenous knowledge. A new applied anthropology, *Current anthropology*, 39(2): 223-52.

SILLITOE, P. (2002) Participant observation to participatory development: Making anthropology work, in Sillitoe, P., A. Bicker and J. Pottier (eds.) *Participating in development: Approaches to indigenous knowledge, ASA*

Monographs 39, London: Routledge, New York:: Taylor and Francis Group, pp. 1-23.

SMITH, C. (1985) The disappearance of another Harare land mark, *Heritage*, 5: 91-4.

SPERBER, D. (1975) *Rethinking symbolism*, London: Cambridge University Press.

SPIERENBURG, M. (2004) *Strangers, spirits and land reforms. Conflicts about land in Dande, Northern Zimbabwe*, Leiden: Brill.

STEELE, M.C. (1972) *Interview with Henry Roger George Howman, M.B.E.*, Place of interview – Interviewee's office, Coghlan building, 4th street, Salisbury, National archives.

STRATHERN, M. (1985) Knowing power and being equivocal: Three Melanesian contexts, in Fardon, R. (ed.) *Power and knowledge. Anthropological and sociological approaches*, Edinburgh: Scottish Academic Press.

STUIVER, M., C. LEEUWIS and J.D. VAN DER PLOEG (2004) The power of experience: Farmer's knowledge and sustainable innovation in agriculture, in Wiskerke, J.C. and J.D. van der Ploeg (eds.) *Seeds of transition: Essays on novelty production niches and regimes in Africa*, Assen: Royal Van Gorcum.

TAMBIAH, S.J. (1990) *Magic, science and the scope of rationality*, Cambridge (UK): Cambridge University Press.

TURNER, B.S. (1985) The practice of rationality: Michel Foucault, medical history and sociological theory, in Fardon, R. (ed.) *Power and knowledge. Anthropological and sociological approaches*, Edinburgh: Scottish Academic Press.

VELSEN, T. van (1967) The extended case method and situational analysis, in Epstein, A. (ed.) *The craft of social anthropology*, London: Tavistock.

VERDON, M. (1998) *Rethinking households: An atomistic perspective on European living arrangements*, London: Routledge.

VIVIAN, J. (1994) NGOs and sustainable development in Zimbabwe. No magic bullets, *Development and Change*, 25(1): 167-94.

WALTHALL, A. (1982) Narratives of peasant uprisings in Japan, *Journal of Asian Studies,* 42(3).

WARREN, M.D., L. Slikkerveer, D. Brokenshaw and W. Dechering (eds.) *The cultural dimensions of development: Indigenous knowledge systems*, UK: Intermediate Technology Publications.

WATERHOUSE, R. and C. VIJFHUIZEN (eds.) (2001) *Strategic women gainful men: Gender, land and natural resources in a different rural context in Mozambique*, Mozambique: University of Eduardo Mondlane, pp. 5-30.

WEINER, D., S. MOYO, B. MUNSLOW and P. O'KEEFE (1985) Land use and agricultural productivity in Zimbabwe, *The Journal of Morden African Studies*, 23(2) 251-85.

WEINRICH, A.K.H. (1983) *African marriage in Zimbabwe and the impact of Christianity*, Gweru: Mambo Press

WEINRICH, A.K.H. (1975) *African farmer in Rhodesia. Old and new peasant communities in Karanga land*, Oxford: Oxford University Press.

WEINRICH, A.K.H. (1973) *Black and white elites in rural Rhodesia*, Manchester (USA): Manchester University Press.

WICKHAM, G. (1986) Power and power analysis beyond Foucault, in Gane, M. (ed.) *Towards a critique of Foucault.*, London: Routledge/Kegan-Paul.

WISKERKE, J.C. and J.D. VAN DER PLOEG (2004) *Seeds of transition: Essays on novelty production niches and regimes in Africa*, Assen: Royal Van Gorcum.

WORBY, E. (1995) Tyranny, parody, and ethnic polarity: Ritual engagements with the state in northwestern Zimbabwe, *Journal of Southern Africa*, 24: 561-79.

WOOLGAR, S. (1983) Irony in the social study of science, in Knorr-Cetina, K.D. and M. Mulkay (eds.) *Science observed. Perspectives on the social study of science*, London: Sage

YUDELMAN, M. (1964) *Africans on the land. Economic problems of African agricultural development in southern, central and eastern Africa with special reference to southern Rhodesia*, Cambridge (USA): Harvard University Press.

ZAWE, C. (forthcoming) *Policy models for irrigation management reform*, Wageningen, PhD Thesis

ZAWE, C. (2000) *Coping with cooperation and maintenance at farmer managed sprinkler irrigation schemes: A case of irrigation management turnover experiment at Musengezi Schemes*, Wageningen: Department of Environmental Sciences, MSc Thesis.

ZEITLYN, D. (1991) Do Mambila cockrels lay eggs? Reflections on knowledge and belief, *JASO*, 22(1): 59-64.

ZINYAMA, L.M. (1995) Sustainability of smallholder food production systems in Southern Africa: The case of Zimbabwe, in Binns, T. (ed.) *People and environment in Africa*, New York: John Wiley and sons.

ZINYAMA, L.M. (1992) Local farm organisations and rural development in Zimbabwe, in Taylor, F. and Mackenzie, F. (eds.) *Development from within: Survival in rural Africa*, London: Routledge.

ZVOBWO, C.J.M. (1996) A history of Christian mission in Zimbabwe 1890-1939, Gweru: Mambo Press.

Index

www.ingramcontent.com/pod-product-compliance
Lightning Source LLC
Chambersburg PA
CBHW061207240926
51846CB00006B/1132
9789004161689